EUROPEAN RIGHT-OF-WAY AND UTILITIES BEST PRACTICES

PREPARED BY THE INTERNATIONAL SCANNING STUDY TEAM

Richard Moeller
FHWA
Team Leader

Joachim Pestinger
Washington State DOT
Team Leader

Myron Frierson
Michigan DOT

Wayne Kennedy
International Right-of-Way
Association

Adele McCormick
Report Facilitator

Catherine Colan Muth
O.R. Colan Associates, Inc.

Janet Myers
Maine DOT

Paul Scott
FHWA

Stuart Waymack
Virginia DOT

and
American Trade Initiatives, Inc.
&
Avalon Integrated Services, Inc.
for the
Federal Highway Administration
U.S. Department of Transportation

American Association of State Highway and Transportation Officials

National Cooperative Highway Research Program
(Panel 20-36)
of the
Transportation Research Board

August 2002

ACKNOWLEDGMENTS

Many people contributed to the success of the Right-of-Way and Utilities Scanning Study. Above all, the team would like to thank the members of the host delegations who so willingly gave their time, resources, and hospitality to make us welcome and provide the team with a wealth of valuable information on right-of-way and utilities practices in their countries. Although too numerous to list here, the individuals the team met with are listed in Appendix C. In addition to those listed, the team would like to thank everyone in the host countries who worked on behind-the-scenes logistics. We appreciate their valuable contribution to the success of this study. Special thanks go also to the superb interpreters who enabled team members to focus their attention on the content of each presentation with such ease.

The team would like to express its gratitude to the staff members of American Trade Initiatives, Inc., (ATI) for their phenomenal efforts, without which this trip would not have been possible. ATI, contracting to the Federal Highway Administration (FHWA), handled all logistical aspects of the study, from pre-trip planning to preparation of this report. The staff's guidance was invaluable. In particular, the team would like to recognize:

- Joe Conn for his expert guidance and assistance in organizing the study.
- John Amborg for his expert leadership and guidance as our "mother duck" and escort during the study.
- Alexandra Doumani for her unfailing assistance and expertise with travel arrangements and funds disbursement.
- Betty Dillon for her assistance in preparing this report for publication.

Finally, this trip would not have been possible without the support and funding of the FHWA Office of International Programs. The team especially would like to thank Donald Symmes and Hana Maier for sponsoring the trip and allowing the team to learn firsthand the right-of-way and utilities best practices used in the European countries visited.

FHWA INTERNATIONAL TECHNOLOGY EXCHANGE PROGRAMS

The Federal Highway Administration's (FHWA) international programs focus on meeting the growing demands of its partners at the Federal, State, and local levels for access to information on state-of-the-art technology and the best practices used worldwide. While FHWA is considered a world leader in highway transportation, the domestic highway community is interested in advanced technologies being developed by other countries, as well as innovative organizational and financing techniques used by FHWA's international counterparts.

The International Technology Scanning Program accesses and evaluates foreign technologies and innovations that could significantly benefit United States highway transportation systems. Access to foreign innovations is strengthened by U.S. participation on the technical committees of international highway organizations and through bilateral technical exchange agreements with selected nations. The program is undertaken cooperatively with the American Association of State Highway and Transportation Officials (AASHTO) and its Select Committee on International Activities, and the Transportation Research Board's (TRB) National Highway Research Cooperative Program (Panel 20-36), the private sector, and academia.

FHWA and its partners jointly determine priority topic areas. Teams of specialists in the specific areas of expertise being investigated are formed and sent to countries where significant advances and innovations have been made in technology, management practices, organizational structure, program delivery, and financing. Teams usually include Federal and State highway officials, private sector and industry association representatives, and members of the academic community.

FHWA has organized about 50 of these reviews and disseminated results nationwide. Topics have included pavements, bridge construction and maintenance, contracting, intermodal transport, organizational management, winter road maintenance, safety, intelligent transportation systems, planning, and policy. Findings are recommended for follow-up with further research and pilot or demonstration projects to verify adaptability to the United States. Information about the scan findings and results of pilot programs are then disseminated nationally to State and local highway and transportation officials and the private sector for implementation.

This program has resulted in significant improvements and savings in road program technologies and practices throughout the United States, particularly in the areas of structures, pavements, safety, and winter road maintenance. Joint research and technology-sharing projects have also been launched with international counterparts, further conserving resources and advancing the state of the art.

For a complete list of International Technology Scanning topics, and to order free copies of the reports, please see pages iv-v.

Website: www.international.fhwa.dot.gov
E-Mail: international@fhwa.dot.gov

FHWA INTERNATIONAL TECHNOLOGY EXCHANGE REPORTS

International Technology Scanning Program: Bringing Global Innovations to U.S. Highways

INFRASTRUCTURE

Geotechnical Engineering Practices in Canada and Europe
Geotechnology—Soil Nailing
International Contract Administration Techniques for Quality Enhancement-CATQEST

PAVEMENTS

European Asphalt Technology
European Concrete Technology
South African Pavement Technology
Highway/Commercial Vehicle Interaction
Recycled Materials in European Highway Environments

BRIDGES

European Bridge Structures
Asian Bridge Structures
Bridge Maintenance Coatings
European Practices for Bridge Scour and Stream Instability Countermeasures
Advanced Composites in Bridges in Europe and Japan
Steel Bridge Fabrication Technologies in Europe and Japan
Performance of Concrete Segmental and Cable-Stayed Bridges in Europe

PLANNING AND ENVIRONMENT

European Intermodal Programs: Planning, Policy and Technology
National Travel Surveys
Recycled Materials in European Highway Environments
Geometric Design Practices for European Roads
Sustainable Transportation Practices in Europe
Wildlife Habitat Connectivity Across European Highways
European Right-of-Way and Utilities Best Practices

SAFETY

Pedestrian and Bicycle Safety in England, Germany and the Netherlands
Speed Management and Enforcement Technology: Europe & Australia
Safety Management Practices in Japan, Australia, and New Zealand
Road Safety Audits—Final Report
Road Safety Audits—Case Studies

Innovative Traffic Control Technology & Practice in Europe
Commercial Vehicle Safety Technology & Practice in Europe
Methods and Procedures to Reduce Motorist Delays in European Work Zones

OPERATIONS

Advanced Transportation Technology
European Traffic Monitoring
Traffic Management and Traveler Information Systems
European Winter Service Technology
Snowbreak Forest Book – Highway Snowstorm Countermeasure Manual (*Translated from Japanese*)
European Road Lighting Technologies
Freight Transportation: The European Market

POLICY & INFORMATION

Emerging Models for Delivering Transportation Programs and Services
Acquiring Highway Transportation Information from Abroad—Handbook
Acquiring Highway Transportation Information from Abroad—Final Report
International Guide to Highway Transportation Information

All publications are available on the internet at www.international.fhwa.dot.gov

CONTENTS

EXECUTIVE SUMMARY ix
    Appraisal and Acquisition x
    Compensation and Relocation x
    Training xi
    Utilities xi
    Project Development xii
    Recommendations and Implementation Strategies xiii
        Appraisal and Acquisition xiii
        Compensation and Relocation xiv
        Training xiv
        Utilities xiv
        Project Development xv
    Ongoing Implementation Activities xvi

CHAPTER ONE – INTRODUCTION 1
    Scanning Study Planning 1
    Team Members 2
    Meetings 2
    Amplifying Questions 3
    Itinerary 3
    Host Delegations 4
    Report Organization 4

CHAPTER TWO – APPRAISAL AND ACQUISITION 5
    Primary Findings 5
        Early Involvement of Property Owners in Design Process 6
        Property Owner Interviews 6
        Limited Use of Appraisal Reviews 6
        Appraisal and Negotiation Functions Performed by Same Person 7
        Project-Specific Legislation to Facilitate Real Estate Phase 7
        Mediation and Quick Payment Processes 7
    Other Observations 7
        Appraisal and Acquisition Responsibilities 7
        Valuation 7

CHAPTER THREE – COMPENSATION AND RELOCATION ASSISTANCE 9
    Primary Findings 9
        Compensation for Acquisition of Right-of-Way 9
        Relocation Assistance 10
        Compensation Claims from Outside Project 11
        Land Consolidation 11

CHAPTER FOUR – TRAINING 13
    Primary Findings 13
        Norway's Five-Year Degree Program 13
        The Netherlands' In-House Training Programs 14
        England's In-House Training Programs 15

Other Observations 15
    German Association for Appraisers 15
    Royal Institution of Chartered Surveyors 15
    Appraiser/Relocation Agent 16

CHAPTER FIVE – UTILITIES RELOCATION AND ACCOMMODATION 17
Primary Findings 17
    Cooperation, Coordination, Communication 17
    Locating Utilities Underground 18
    Utility Corridors 18
    Recognizing Pipelines as a Transportation Mode 19
    Avoiding Unnecessary Utility Relocations 20
    Utilities in Design-Build Contracts 20
    Master Utility Agreements 21
Other Observations 21
    Utility Installations by Highway Contractors 21
    Cost Sharing 22
    Right-of-Way Acquisition for Utilities 22
    Damage Prevention 23
    Protected Highway Designation 24
    Minimizing Pavement Cuts 24
    Geographic Information Systems 24
    Fiber Optics and Wireless Communications 25

CHAPTER SIX – PROJECT DEVELOPMENT 26
Primary Findings 26
    Multidisciplinary Team Approach 26
    Design-Build 26
    Multidimensional and Inclusive Planning Processes 27
    Definition of Problems and Solutions 28
    Planning Stage Feasibility Analysis 28
    Land Consolidation 28
    Realistic Right-of-Way Budgets and Solutions 28
    External Communication, Coordination, and Participation 29
    Flexible Early Acquisition Tools 29
    User-Friendly Right-of-Way Plans 30
    Right-of-Way Databases and GIS Systems 31
Other Observations 31

CHAPTER SEVEN – RECOMMENDATIONS 33
Appraisal and Acquisition 33
    Early Involvement of Property Owners 33
    Property Owner Interviews 33
    Limited Use of Appraisal Reviews 33
    Appraisal and Negotiation Functions Performed by Same Person 33
    Incentive Payments 34
Compensation and Relocation 34
    Voluntary Land Consolidation Pilot Program 34

    Business Reestablishment and Relocation Program 34
    Residential Relocations 34
  Training 34
    Pre-Employment and Employee Education and Training 34
    Mentoring Methods 35
  Utilities 35
    Cooperation, Coordination, and Communication 35
    Underground Utilities 35
    Utility Corridors 35
    Recognizing Pipelines as a Transportation Mode 36
    Utilities in Design-Build Contracts 36
    Master Utility Agreements 36
    Additional Recommendations for Utilities 36
  Project Development 38
    Right-of-Way and Utilities Functions in Design-Build Process 38
    Corridor Preservation 38
    Right of Entry and Early Acquisition Methods 39
    Information Clearinghouse on Right-of-Way and Utilities Databases 39

CHAPTER EIGHT – IMPLEMENTATION ACTIVITIES 40
  Right-of-Way Program 40
    Land Consolidation 40
    Right-of-Way Experimental Projects 41
    Right-of-Way Exchanges, Meetings, and Presentations 42
    Future Actions 42
  Utilities Program 43
    Experimental Utility Projects 43
    Utility Research/Technology Transfer Projects 44
    Utility Meetings and Presentations 44
    Future Actions 45
  Right-of-Way and Utilities Guidelines and Best Practices Report 45
    Project Development 45
    Appraisal and Appraisal Review 46
    Acquisition 46
    Training 46

APPENDIX A – TEAM MEMBERS 47
  Team Members 47
  Team Biographies 48

APPENDIX B – AMPLIFYING QUESTIONS 51

APPENDIX C – CONTACTS IN HOST COUNTRIES 55

APPENDIX D – WOODROW WILSON BRIDGE PROJECT REPORT 57

EXECUTIVE SUMMARY

During the past 20 years, highway right-of-way acquisition and utilities accommodation in the United States have become significantly more complex. At the same time, right-of-way and utilities personnel have come under increasing pressure to provide cleared right-of-way more quickly. The Federal Highway Administration (FHWA) developed a National Strategic Plan to enhance communities through highway transportation projects using innovative acquisition of right-of-way, sensitive and effective relocation of affected residences and businesses, and relocation and accommodation of utilities with minimal impact and disruption to the communities.

As part of the American Association of State Highway and Transportation Officials' (AASHTO) strategic plan assignment, the AASHTO Subcommittee on Right-of-Way and Utilities completed a nationwide review of processes and procedures to identify best practices in the United States. The study outlined process improvements in the following areas:

- Right-of-way and utilities involvement in project development
- Property appraisal and appraisal review
- Acquisition of total property rights, easements and permits
- Relocation assistance to owners, tenants, businesses, and farm operations
- Property management of acquired right-of-way
- Utilities coordination, adjustments, and relocations for highway projects
- Management practices for effective right-of-way and utilities operations
- Training programs and mentoring procedures for staff development

In March 2000, FHWA, AASHTO, and the Transportation Research Board (TRB) jointly sponsored an international scanning study to observe right-of-way and utility coordination practices in four European countries. The scanning study delegation identified practices used in the selected countries that, if implemented in the United States, will help ensure timely procurement and clearance of highway right-of-way and adjustment of utilities. The dissemination of information about and potential adoption of European right-of-way and utilities techniques and best practices will enhance the ability of State and local agencies to streamline delivery and improve the quality of right-of-way services.

The U.S. delegation included members representing State departments of transportation (DOTs) in Maine, Michigan, Virginia, and Washington State; FHWA; and the private sector, including representatives from the International Right-of-Way Association (IRWA) and O.R. Colan Associates, Inc. These panel members offered expertise in many right-of-way and utilities activities, including project development, appraisal and appraisal review, acquisition, property management, condemnation, relocation, and utilities coordination and accommodation. Team members are listed in Appendix A.

The team met with transportation officials in Oslo, As, and Moss, Norway; Bonn, Germany; The Hague, Netherlands; and London, England. Host officials provided a wealth of information on right-of-way and utilities practices in their respective countries, as well as insight into similar practices in neighboring countries. Travel between cities and countries afforded the team additional opportunities to observe innovative practices.

Team meetings were held at the beginning, midpoint, and end of the two-week study to share observations and discuss practices identified as having value for potential implementation in the United States.

Findings and observations in this report are grouped into the following chapters:

- Appraisal and Acquisition
- Compensation and Relocation
- Training
- Utilities
- Project Development

Each chapter includes primary findings the team believes have the most significance and/or implementation value. Other observations that may have potential implementation value in the United States also are included.

APPRAISAL AND ACQUISITION

The countries the team visited have an underlying philosophy of sensitivity to the needs of property owners. In some cases, this philosophy replaces the need to have prescriptive regulations on how to conduct appraisal, acquisition, and relocation procedures. The primary findings identified by the team for appraisal and acquisition reflect this sensitivity.

Practices used in these countries encourage property owner involvement before completion of final right-of-way plans and use an extensive property owner interview process. They make a conscientious effort to limit the number of people contacting the property owner, including assigning one person to serve as appraiser and negotiator for acquisition and relocation services.

Highway agencies in these countries reduce the time needed to provide acquisition offers to property owners by limiting the need for appraisal reviews and through passage of special enabling legislation to streamline the acquisition process. They use mediation and quick payment processes to facilitate settlements and payments to property owners. These actions underscore the desire of the highway agencies to provide a fair and equitable method for acquiring right-of-way.

COMPENSATION AND RELOCATION

All of the countries visited have a compensation framework similar to that used in the United States. In many cases, however, compensation includes elements not always compensable in the United States. Compensation includes provisions for payment for land acquired, damages to remaining property, and relocation reimbursements. The impact on properties outside the project limits also is considered.

The countries all provide liberal payments to businesses affected by property acquisition, project construction, or highway operations. These payments range from liquidation and acquisition of businesses to the negotiated reimbursement of moving and relocation costs and incidental expenses incurred by displaced businesses.

The team was intrigued by the land consolidation concept used in Norway, Germany, and the Netherlands. Land consolidation involves adjusting property boundaries in the area of a highway project and redistributing the land to affected landowners. Land for sale next to the project also can be acquired and reassembled with properties divided by the project. The idea is to create better parcels for continued agricultural use or more desirable parcels for development. In Norway, the acquiring agency or one property owner affected by the project can request an investigation into land consolidation. In Germany and the Netherlands, a consensus of landowners is required to begin the land consolidation process.

TRAINING

Training requirements vary among countries, but they all emphasize formal training and continuous employee development. Programs focus on college curriculums leading to a degree in right-of-way and internal training courses, including small workshops and mentoring.

The team visited the Agricultural University of Norway in As, which offers a five-year degree program covering property rights and land law. The university began offering a program in land use planning in 1898, and has offered a curriculum leading to a career in right-of-way and land consolidation since 1960. Most top-ranking staff members in the Norwegian Public Roads Administration's Land Acquisition and Real Estate Division are graduates of this program.

No universities in the other countries the team visited offer a right-of-way curriculum, but each country has qualification requirements for right-of-way staff. Right-of-way personnel the team met with in Germany were trained as economists, attorneys, and engineers. The Netherlands requires a university degree in almost any subject area, and England normally requires entry-level staff to meet the minimum academic qualifications for recruitment to the civil service. Both the Netherlands and England have developed strong in-house training and continuing education programs.

It is universally recognized in the countries visited that, along with appropriate education, a good right-of-way agent must be mature and have strong people skills. In the Netherlands, employees must demonstrate these traits before being deemed eligible for the two-year in-house training for right-of-way agents.

UTILITIES

Team members were impressed with the strategies used in each country for accommodating and relocating utilities on or near highway rights-of-way. Most of the countries make special efforts to enhance relationships between highway and utilities officials by improving coordination, cooperation, and communication. In several countries, jurisdiction-wide master agreements with each utility company are used to avoid having to develop new utility agreements for every project. Germany tries to avoid the need to relocate utilities during highway construction through design measures.

Roadside safety in Europe has been greatly improved by placing most utilities underground. Team members were impressed not only by the added safety of underground utilities, but also by the enhanced landscape aesthetics. In the Netherlands, all utilities except high-voltage transmission lines are underground.

The team also was intrigued by the Netherlands' concept of recognizing utilities as a mode of transportation, joining highway, air, water, and rail transportation. Although the Dutch policy focuses on gas pipelines, this concept may have broader applicability in moving other liquids or slurries now transported by truck in the United States.

Several countries have established utility corridors for highway crossings. In some cases they have established corridors for longitudinal installations to consolidate utility locations, maximize use of limited available land, and minimize or eliminate road openings. These corridors may include empty conduit for future installations and joint trenching.

Utilities are included as essential components of design-build contracts in England. This is advantageous to the Highways Agency because it transfers the risk of utility-related delays to the highway contractor, reducing claims for delays and large cost overruns.

PROJECT DEVELOPMENT

Several countries have adopted the project management approach to project development, including the use of multidisciplinary teams. Practices include right-of-way participation that begins at the planning stage, budget and schedule commitments with a sign-off by functional representatives and project managers, and accountability for delivery on those commitments.

England uses design-build practices extensively in its program. Although right-of-way acquisition remains the responsibility of the Highways Agency, officials believe the potential for delegating some or all acquisition activities to design-build contractors merits evaluation.

Each country has an extensive planning process that includes significant input from affected property owners, community members, and local authorities. In several countries, zoning and land use plans prepared at the local or regional level govern decisions about the location of the transportation infrastructure.

During the planning process, the European countries define specifically the problems the project will address and describe how it will achieve intended results. Several countries also perform broad feasibility reviews before acquisition. The delegation noted that all of the countries budget enough time and funding for projects to allow appropriately timed and scoped acquisitions and relocations.

The countries appeared to engage in more extensive public coordination than is typical in the United States. Particularly useful practices are:

- Field reviews by the project manager or designer and right-of-way team member to meet with affected property owners early in the development of the project.

- Encouraging owner participation in design issues at early stages of project development.

Each country has a method for facilitating early possession or acquisition. These methods, including advance payment and right-of-entry, provide a great deal of flexibility.

Each country is developing a system for managing data relevant to right-of-way functions. In addition to project file data management, several use geographic information systems (GIS) technology for tracking all land use, including right-of-way.

Some countries also establish standard right-of-way acquisition limits, such as minimums of one meter from the back slope of the ditch and three meters from the edge of pavement.

RECOMMENDATIONS AND IMPLEMENTATION STRATEGIES

The host countries provided the U.S. delegation with a wealth of information on their right-of-way and utilities practices. The team developed the following list of practices with potential for implementation in the United States to help ensure timely procurement and clearance of highway right-of-way and adjustment of utilities. The findings, observations, conclusions, and recommendations are those of the scanning team and not FHWA.

Appraisal and Acquisition

Early Involvement of Property Owners in Design Process
The scanning team recommends that FHWA encourage States to consult affected property owners before project design is completed to assess the impact of the proposed design and to determine if revisions are warranted. Appropriate use of this practice could result in more timely purchases and reduce damages to affected properties.

Property Owner Interviews
The team recommends that FHWA encourage States to use a more extensive interview process to discuss the project's impact with property owners and gain an understanding of how property owners use their property. Information obtained will be used to determine if further investigation of possible damages is necessary.

Limited Use of Appraisal Reviews
The team recommends that FHWA develop a risk management-based appraisal review system to use on pilot projects in several States. The results will be used to recommend regulatory changes necessary to adopt a risk management-based appraisal review system similar to those used in some European countries. The goal is to determine whether such a system—for instance, auditing a sample, reviewing all complex appraisals, or setting review thresholds—can protect the quality and integrity of the valuation process while saving overall project time and costs.

Appraisal and Negotiation Functions Performed by Same Person
The team recommends that FHWA implement a pilot program allowing several States to use the same agent to conduct both the appraisal and negotiation functions on a parcel.

Compensation and Relocation

Voluntary Land Consolidation Pilot Program
The team recommends that FHWA research the ability of States to accomplish voluntary land consolidation and implement a pilot program to evaluate the benefits.

Business Reestablishment and Relocation
The team suggests that FHWA evaluate items eligible for business reestablishment and relocation reimbursement in the Netherlands and England. The European experience and the results of the recent FHWA-sponsored "Business Payments and Services" research can be used to support changes in Federal legislation and regulations.

Training

Pre-Employment and Employee Education and Training
The team recommends evaluating development of a pre-employment and employee education and training program. This includes exploring the potential for recruiting one or more colleges to provide this service, which would include a degree program in right-of-way careers and a continuing education program using distance-learning techniques. This proposal expands on the possibility of the Federal Government establishing an academy for real estate services.

A panel of representatives from FHWA, IRWA, AASHTO, and a private consultant will pursue this training concept. FHWA will act as the lead to contact colleges and online learning centers, with the goal of developing and implementing such a curriculum by Fall 2002.

Mentoring Methods
The team suggests that FHWA evaluate mentoring activities in each State through AASHTO's Internet site, summarize mentoring methods in the United States and Europe, and recommend adoption to the States.

Utilities

Cooperation, Coordination, and Communication
The team recommends that FHWA and AASHTO's Subcommittee on Right-of-Way and Utilities encourage State DOTs to enhance cooperation, coordination, and communication with utility companies.

Underground Utilities
The team recommends that State DOTs continue to develop or enhance utility pole safety programs, including considering underground utilities as a possible countermeasure. Although locating utilities underground can be costly, States are encouraged to give appropriate weight to factors such as safety, environmental impact, and community effects in their decision-making process.

Utility Corridors
The team recommends that State DOTs consider establishing utility corridors, including placing conduit for future use and using joint trenching techniques, and requiring utility companies to coordinate installation of facilities within these corridors.

Recognize Pipelines as a Transportation Mode
The team suggests that State DOTs encourage use of pipelines as a transportation mode by facilitating research and developing methods to exploit pipeline transport. This may include establishing routes and corridors for pipeline companies or funding construction and operation of pipelines.

Avoiding Unnecessary Utility Relocations
The team recommends that State DOTs avoid unnecessary relocation of utilities during highway construction by identifying all utilities early in the project development process and designing around them wherever possible.

Utilities in Design-Build Contracts
The team suggests that States not already doing so consider including utilities in contracts for design-build projects, thereby transferring risks of utility-related delays to highway contractors.

Master Utility Agreements
The AASHTO Subcommittee on Right-of-Way and Utilities has established master agreements as a best practice to eliminate the need for approvals on each individual contract. The team recommends that States not already using master utility agreements consider doing so. AASHTO and/or FHWA should consider developing model master agreements or distributing sample master agreements from States that use them.

Project Development

Incorporate Right-of-Way and Utilities Functions in Design-Build Process
The team suggests that FHWA and AASHTO continue to encourage State right-of-way and utilities personnel to study the advantages of design-build contracting, which include shortening the project development process by eliminating many procedural procurement processes.

Corridor Preservation
The team recommends that FHWA initiate a work group to reevaluate methods for corridor preservation and create one or more pilot projects to test corridor preservation and land consolidation techniques.

Rights of Entry and Early Acquisition Methods
The team believes FHWA and State DOTs should evaluate methods for rights of entry and early acquisition to facilitate early entry onto property for project construction. They should consider expanding these methods by using risk management concepts, while ensuring that property owner rights are protected.

Information Clearinghouse on Right-of-Way and Utilities Databases
The team encourages the AASHTO Subcommittee on Right-of-Way and Utilities to establish an information clearinghouse on right-of-way and utilities databases, including GIS, for project development, tracking, and management.

ONGOING IMPLEMENTATION ACTIVITIES

Implementation efforts are under way on many of the scanning team's recommendations. Some activities of special interest are noted below. A summary of implementation activities is included in Chapter Eight.

In December 2000, FHWA issued a national policy on land consolidation. Several States along the proposed Interstate 69 route indicated interest in this concept, and Mississippi is actively considering using the land consolidation policy.

The Virginia DOT, in cooperation with FHWA, used an experimental tenant relocation incentive program to maintain the project schedule on the new Woodrow Wilson Bridge construction. This program allowed successful relocation of tenants from four buildings in time for subsequent construction activities to continue as planned. The Virginia DOT Woodrow Wilson Bridge Project report entitled "Cost and Schedule Savings from the Early Move Incentive Program for the Hunting Tower and Terrace Buildings" is included in Appendix D.

Initiatives are under way with Morgan State University in Baltimore, Maryland; Delaware Technical and Community College in Stanton, Delaware; and Marylhurst University in Portland, Oregon, to develop right-of-way training, including degree programs and capabilities for distance learning.

The Virginia DOT began a pilot program in September 2000 to determine the feasibility of paying preliminary engineering costs for all utility relocations. Early indications are that the benefits of this practice outweigh the cost.

The North Carolina State University Center for Transportation and the Environment initiated a literature search on the feasibility of recognizing pipelines as a mode of transportation. The center found that the Texas Transportation Institute has designed a system in which pipelines would be used to carry freight from Dallas, Texas, to Laredo, Mexico. The institute is looking for a funding source to build a prototype.

Chapter One
INTRODUCTION

Highway right-of-way acquisition and utilities accommodation in the United States has become significantly more complex during the past 20 years. At the same time, right-of-way and utilities personnel are under increasing pressure to provide cleared right-of-way more quickly. Included in the Federal Highway Administration's (FHWA) National Strategic Plan are goals to enhance communities through highway transportation projects using innovative right-of-way acquisition, relocation of affected residences and businesses, and relocation and accommodation of utilities with minimal disruption to communities.

As part of the American Association of State Highway and Transportation Officials' (AASHTO) strategic plan assignment, the AASHTO Right-of-Way and Utilities Subcommittee conducted a nationwide review of processes and procedures to identify best practices. This study addressed guidelines and best practices in project development, appraisal and appraisal review, acquisition, relocation, property management, utilities, management practices, and training. The "Right-of-Way and Utilities Guidelines and Best Practices" draft was published in January 2000.

While the right-of-way study was being conducted in the United States, FHWA, AASHTO, and the Transportation Research Board (TRB) chose the topic to study on an international level as well. The International Technology Scanning Program, coordinated by FHWA's Office of International Programs, provides a means to identify, assess, and import foreign highway technologies and practices for implementation in the United States.

A scanning team of right-of-way and utilities experts from Federal, State, and private sector agencies was formed. In March 2000, the team traveled to Norway, Germany, England, and the Netherlands to observe right-of-way and utilities best practices and identify techniques with significant potential value for implementation in the United States.

This report describes the team's findings and observations. It includes their recommendations for potential adoption of European right-of-way and utilities techniques and best practices that will enhance the ability of State and local agencies to streamline delivery and improve the quality of right-of-way services.

SCANNING STUDY PLANNING

A Joint AASHTO-FHWA International Highway Technology Scanning Program was established in 1998. This program includes joint proposal and selection of study topics and joint responsibility for U.S. implementation of useful techniques and practices identified during scanning studies. Twelve scanning studies every two years are conducted under this program. Co-chairs from AASHTO and FHWA head each scanning team and recruit other team members. Teams typically include members from FHWA, State highway departments, industry associations, the private sector, and academia.

Once FHWA and AASHTO identify the need for a scanning study and a team is formed, the team meets to address key aspects of the trip and develop amplifying questions. The amplifying questions are submitted to transportation officials in the host countries to

CHAPTER 1

give them an idea of the scope of information the team is seeking. The amplifying questions for this scanning study are included in Appendix B.

TEAM MEMBERS

Team members for the Right-of-Way and Utilities Scanning Study represented FHWA, State departments of transportation (DOTs), and the private sector. Co-chairs were Richard Moeller, FHWA, and Joachim Pestinger, AASHTO. Four team members represented State DOTs: Myron Frierson of Michigan, Janet Myers of Maine, Pestinger of Washington State, and Stuart Waymack of Virginia. Private sector members included Wayne Kennedy of the International Right-of-Way Association (IRWA) and Catherine Muth from O.R. Colan Associates, Inc. Moeller and Paul Scott represented FHWA. Appendix A contains biographies of team members.

Figure 1. Scanning team members include (left to right) Dick Moeller, Joe Pestinger, Adele McCormick, Cathy Muth, Stuart Waymack (behind), Myron Frierson, Janet Myers, Wayne

MEETINGS

American Trade Initiatives, Inc., which provides logistical support and guidance for scanning studies, scheduled four team meetings at various stages of the right-of-way study. At the first meeting in Washington, D.C., team members established the study's focus, discussed problem areas in the right-of-way and utilities accommodation process, and developed amplifying questions.

The second team meeting was held in Oslo, Norway, before the first day of presentations by the team's Norwegian hosts. This meeting provided an opportunity for the team to review study objectives and the host countries' agendas.

The third meeting was held halfway through the trip in The Hague, Netherlands, to allow the team to refocus and regroup to ensure the scanning study's objectives were being met. This was a pivotal meeting at which the team discussed the first week's findings and observations and set the course for the second week.

The final meeting was held in London, England, on the last day of the study. The team used this meeting to review findings and develop recommendations for implementation.

AMPLIFYING QUESTIONS

The team developed amplifying questions to give their European hosts an understanding of the scope of information the team desired. Questions were divided into nine categories:

1. Right-of-way and utility laws, regulations, policies, and programs
2. Right-of-way and utility involvement in project development
3. Property appraisal and appraisal review
4. Acquisition of property rights, easements, and permits
5. Relocation assistance to owners, tenants, businesses, and farm operations
6. Utility coordination, adjustments, and relocation for highway projects
7. Public involvement
8. Property management of real estate acquired for highway right-of-way
9. Training programs and mentoring procedures for right-of-way staff

Each topic included specific questions intended to provide the team with an understanding of the host country's right-of-way and utilities practices and techniques. Amplifying questions are listed in Appendix B.

ITINERARY

The Right-of-Way and Utilities Scanning Study took place in March 2000. The team met with highway right-of-way and utilities experts in Oslo, As, and Moss, Norway; Bonn, Germany; The Hague, Netherlands; and London, England. Two bus trips—one from Berlin to Bonn and another from Bonn to The Hague—allowed the team to experience highway right-of-way and utilities operations in those countries firsthand. The team's Norwegian hosts organized another bus trip that took the team through the new Oslo Fjord Tunnel, the world's longest sub-sea tunnel for car traffic, and to other points of interest, including the Agricultural University of Norway in As and the Ostfold County Road Office in Moss.

Table 1. Right-of-Way and Utilities Team Meetings

LOCATION	DATE	PURPOSE
Washington, D.C.	December 1999 (3 months before study)	Set study focus and develop amplifying questions
Oslo, Norway	March 12, 2000 (start of study)	Review team objectives and host country agendas
The Hague, Netherlands	March 19, 2000 (mid-study)	Discuss findings and review objectives
London, England	March 25, 2000 (final day of study)	Identify key findings and develop recommendations for implementation

CHAPTER 1

HOST DELEGATIONS

The team met with representatives from the Norwegian Public Roads Administration (NPRA) and the Agricultural University of Norway; Bundesministerium fur Verkehr (Federal Ministry of Transport) in Germany; Ministerie van Verkeer en Waterstaat, Rijkswaterstaat (Ministry of Transport, Public Works, and Water Management) in the Netherlands; and the Highways Agency and Valuation Office in England. Appendix C lists members of the host nations' delegations. These host officials provided the team with a wealth of information on their nations' right-of-way and utilities techniques and practices. The team was grateful for the time and energy the host delegations expended to make their visit informative and rewarding.

REPORT ORGANIZATION

During meetings in each host country, the team identified many interesting right-of-way and utilities practices that may be of value for implementation in the United States. The following chapters present these findings in the categories of appraisal and acquisition, compensation and relocation, training, utilities, and project development. The chapters include findings the team believes have the greatest potential for U.S. implementation. Other observations include items that may have implementation value in specific instances. Recommendations and implementation strategies are included in the final chapters.

Figure 2. The Right-of-Way and Utilities Scanning Team meets at St. Christopher House in London.

Chapter Two
APPRAISAL AND ACQUISITION

The team was surprised to find so much similarity between right-of-way acquisition processes in the four European countries and those in the United States. While their governments and cultures differ, the countries share basic principles that guide the process. The right-of-way professionals the team met expressed feelings of privilege to work in a profession that allows them to participate in building their countries' future and help individuals who must relocate during the process.

In all of the countries visited, team members found an underlying philosophy of sensitivity to the needs of property owners. In some cases, this philosophy replaces the need for regulations on how to conduct appraisal and acquisition processes.

Figure 3. The Norwegian highway agency built a bridge to allow safe access to a farm and preserve it for the community's benefit.

In Norway, the team visited a small farm next to a new highway. The highway agency had provided a bridge to allow children to continue to travel safely to the farm for horseback riding lessons. It would have been less expensive to buy the farm than to build the bridge, but the bridge was an example of cooperation between local and national planners and their desire to preserve the farm for the community's benefit. In general, the team found that in all countries visited the sense of community welfare is stronger than it is in the United States.

Team members observed several appraisal and acquisition practices with implementation value in the United States. These techniques include early involvement of property owners in the design process, property owner interviews, limited use of appraisal reviews, appraisal and negotiation functions performed by the same person, project-specific legislation to facilitate the real estate phase, and mediation and quick payment processes.

PRIMARY FINDINGS

In the countries visited, practices used to provide for more property owner input included early involvement of property owners in the design process and an extensive property owner interview process. Allowing one person to serve as appraiser and negotiator reduced the number of people contacting the property owner. Reducing the need for appraisal reviews and passage of special legislation allowed property owners to receive more timely offers. These actions underscored the objective of providing a fair and equitable method for acquiring right-of-way.

Early Involvement of Property Owners in Design Process

Affected property owners are consulted before completion of project design to assess the impact of the proposed design and to determine if a revision is warranted. Early consultation with property owners provides more information to designers to incorporate in project design decisions. Use of this practice in the countries visited resulted in more timely purchases and reduced damages to affected properties. Early and frequent contact with property owners is considered essential to successful acquisition.

In Norway, the right-of-way agent in charge of a project conducts outreach to assure that the public knows what is going on and understands the process. For large projects, Norway prepares environmental impact studies modeled after those used in the United States.

In the Netherlands, the process for right-of-way acquisition is considered during the project planning stage. Estimates for the cost of right-of-way are provided during the design stage.

In England, the Highways Agency schedules a public forum before a route is decided to let property owners know what is being considered. Early involvement with property owners helps the agency identify ways to mitigate damages. For example, an underpass or overpass may be provided if a project splits a farmer's field or the agency may delay work until a farmer has harvested his crop.

Property Owner Interviews

Acquisition staff members meet with property owners at length to gain an understanding of how they use the property and discuss a project's impact. This information is used to determine if further investigation of possible damages is necessary. If so, appropriate experts are assigned to assess the project's impact on the property. Findings from the in-depth interview, appraisal, and expert analysis contribute to development of a comprehensive estimate of just compensation, which facilitates negotiations with the property owner.

Limited Use of Appraisal Reviews

England, Norway, and the Netherlands use the appraisal review process on a limited basis. In England, instead of preparing a formal review of every appraisal, staff members review appraisals by sampling. In Norway and the Netherlands, appraisals are reviewed informally. Questions and concerns are addressed through conversations with the appraiser, but no formal written appraisal review is prepared. Appraisal staff members in these countries believe this saves time and money while maintaining an acceptable level of quality.

In the Netherlands, the appraisal review centers on whether the appraiser has done an adequate job supporting the estimate of value. Most reviews are brief and limited in scope.

In England, most appraisals are not reviewed because the Highways Agency believes that if it hires competent professionals there should be no need to review their work.

Appraisers, called surveyors, conduct negotiations as well. Private appraisers also do valuations and negotiations.

Appraisal and Negotiation Functions Performed by Same Person

In Norway and England, the person performing the appraisal often handles negotiations and relocation activities also. All the countries visited use a single agent for negotiation and relocation activities to maintain good rapport between the property owner and the agency and to ensure consistency in communication.

Project-Specific Legislation to Facilitate Real Estate Phase

Several countries have adopted special legislation to address project development and delivery issues, including right-of-way. The objective of this type of legislation is to accelerate completion of transportation improvement projects. Examples observed provided for fewer reviews and permits, shortening process times. These practices, used through special statutes, are being evaluated to determine if they can be expanded to normal operations.

Mediation and Quick Payment Processes

Use of mediation and quick payment processes facilitates settlements and payments to property owners. These actions underscore the highway agencies' desire to provide a fair and equitable method for acquiring right-of-way.

OTHER OBSERVATIONS

Appraisal and Acquisition Responsibilities

In Norway, NPRA's Land Acquisition and Real Estate Division deals only with national and county roads. Municipalities take care of municipal roads. NPRA acts as consultant in land acquisition matters and legal questions in connection with planning and handles right-of-way planning, appraisals, and acquisition.

In Germany, the Federal Ministry of Finance has responsibility for land acquisition and is responsible for setting compensation and finding replacement areas, if they are not under the management of the individual construction agency.

Normally, Germany's road administration commissions outside experts to appraise forestry and agricultural areas, as well as buildings and other structures. Outside experts perform all commercial property appraisals. A property owner may hire an appraiser and have reasonable costs reimbursed by the government. If rural property is involved, a government appraiser will prepare the appraisal. If road construction will impair the rest of the property or the acquisition of right-of-way will leave part in an unusable condition, damages will be paid to the owner or the entire property will be acquired.

Valuation

NPRA's process allows real estate officials to negotiate with or without an appraisal, using the principle of equality to determine the value from other properties sold. This is essentially the comparison approach used in the United States. The phrase "fair market value" is used in Norway, but it is defined as the value the buyer wishes to pay for the property. When valuing commercial property, appraisers compute value based on the capitalization of future net income. They are required to pay the higher of the two approaches to value. Offers to acquire are made in person and then submitted in writing, although sometimes only a telephone call or letter is required, depending on the individuals involved.

Norway has a national registry of all ownerships with a current fair market value assigned. This information is readily available to all property owners, and the result is a property value acceptable to owners and public officials alike. When determining the value of land acquired, basic appraisal principles are applied and before-and-after values are estimated. Full fee title is acquired, except that it does not include subsurface sand, gravel, etc. Owners may choose to have an attorney present. If it is necessary to go to court, the court sets the attorney's fee for one day in court and two days of planning. An owner can make a request for an engineer and, if the NPRA agrees, the owner can select an engineer and the NPRA will pay the fee. The project manager makes it clear, however, that payment will only be made for good work. The right-of-way agent is the contact between the owner and the NPRA.

If land is acquired from another agency, the NPRA does not have to pay for it unless it is used for a commercial enterprise. Excess property is sold, but it is not leased unless it is leased to someone who works for the NPRA. In such cases, fair market rental must be paid.

In the Netherlands, the government will pay the highest price that can be supported by the appraisal for acquisition of full property rights. An appraisal is required for every acquisition and contract appraisers are used. In general, a civil law judge decides disputes about property value. Mediation or arbitration is not used. The property owner is entitled to hire an appraiser and the government will pay for that appraisal.

In England, the basic principal of compensation is the land's market value. There are a number of statutes and precedents from earlier cases. Disputed claims can be referred to the Lands Tribunal, which deals with valuation matters. Government valuers, or appraisers, in England's Valuation Office can operate on behalf of any government body. They work for the highway agency on such matters as compulsory purchase and disposal, but the bulk of their work is for taxation purposes.

In several countries, noise walls are used extensively and payments can be made for installation of noise insulation.

Chapter Three
COMPENSATION AND RELOCATION ASSISTANCE

The European compensation model is broader than the one we know in the United States. The compensation package in the four European countries visited includes provisions for payment for land acquired, damages to remaining property, and relocation reimbursements. In addition, greater consideration is given to reimbursement of claims for impacts to properties situated outside the project.

PRIMARY FINDINGS

Compensation for Acquisition of Right-of-Way

All of the countries visited have procedures in place to compensate property owners for acquisition of private property for public purposes. These processes vary from country to country, depending on laws and customary practices.

This is similar to the United States, where compensation depends on the laws of the State in which a State or local agency is acquiring property. Alternatively, the U.S. Government acquires property according to Federal laws and regulations.

A common denominator in both European and U.S. compensation models is estimation of the fair market value of the property being acquired. This concept serves as the basic criterion for formulating a fair offer to the property owner for negotiation purposes. In the United States and Europe, laws, regulations, customary practice, and implementing procedures further refine the concepts of fair market value and just compensation, creating unique situations in each country.

What is compensable is an issue in both the United States and Europe. There is little debate in the United States that compensation is in order when there is a direct physical taking of property. European models are similar, although some go much further. Nuances that come into play in handling the effects of an acquisition include:

- Damages
- Whether and how benefits may offset just compensation for the land acquired
- Before-and-after appraisal procedures
- Whether consequential damages are actionable under governing law
- Business damages
- Relocation assistance
- Definition of fair market value
- Physical impacts to property outside the project

In the United States, Federal laws pertaining to federally assisted property acquisitions by States or local public agencies recognize and allow differences among various State laws, but require States and local public agencies to pay compensation based on both Federal and State constitutional requirements.

The United States does not have national uniformity in how just compensation is calculated or paid, except that the property must be appraised and an offer made based on the concept of fair market value. While all States will meet the constitutional requirement of just compensation, some States also will compensate for other damage elements.

Relocation Assistance

Compared to acquisition compensation, relocation assistance services and benefits are more uniform in the United States because most States did not have such requirements before 1971. At that time, a Federal law was enacted that requires a minimum threshold of benefits and services for projects involving Federal funding. All States now administer programs that comply with Federal requirements when Federal funds are involved. Most States follow the same procedures when Federal funds are not involved, although some have a dual standard that allows them to operate differently if no Federal funding is involved. In addition, some States have enacted relocation assistance laws and programs that go beyond Federal requirements.

All the European countries visited provide relocation assistance to those displaced by the government's acquisition of property. With the exception of England, programs are less structured than those in the United States, but all generally achieve the same objective of helping the displaced family, business, or farm relocate successfully. England's program, though more structured than the others, also has elements that are general in nature.

In Norway, relocatees must find their own replacement housing, but they are given reimbursement for the cost of relocating. If they relocate from substandard housing, they are reimbursed for relocation into standard housing. If there is agreement on the value of the property taken but not on relocation costs, the owner can go to court for a determination of what reimbursement should be.

If property acquisition results in an owner having to relocate, the German government does not have to pay the difference between the value of the property taken and the price of replacement property. Programs are available, however, to pay supplements and cover the cost of interest payments on loans taken out to buy new property.

The Netherlands has no limit on relocation expenses, although it does limit compensation for future damages. The underlying premise is that the government must put relocatees in as good a position and condition regarding value as they were in before the acquisition. Relocation costs are included in the appraisal for determining compensation. The burden of proof of damages rests with the government. All the owner has to do is file a damage claim and a panel will review the claim and decide the acceptable amount. The panel is composed of real estate managers, attorneys, and accountants who do not work for the government. The panel's findings must be as judgment-proof as possible, so the government's lists are large and panelists are chosen carefully. The same process is used to reimburse costs for consultants.

England's legal framework offers limited possibilities for government relocation advisory services, but some relocation needs can be accommodated during the negotiation process. In some case, relocatees find it difficult to obtain necessary permits for a new location.

CHAPTER 3

Compensation Claims from Outside Project

In some countries, compensation extends beyond right-of-way limits to properties outside the project adversely affected by construction or operation of the facility. In the United States, most States and the Federal Government have refrained from offering compensation in such cases, although a trend appears to be developing toward greater sensitivity to environmental and other impacts created by highway projects.

An example is a new concept called context-sensitive design, which addresses various concerns that arise when a new highway facility is planned and designed. The process can include consideration of impact issues created by a project on properties next to it. FHWA has issued an advance notice of proposed rulemaking to solicit comments on whether insulation of noise-impacted structures should be considered routinely in project development.

The Netherlands will compensate for business interruption in the case of an expropriation, as well as for the cost of noise reduction. These are also compensable items in England.

Land Consolidation

In all of the countries visited except England, the concept of land consolidation is used occasionally in conjunction with right-of-way acquisition. Norway's Land Consolidation Act of 1821-1859 has been amended often, most recently in 1978. Land acquired for consolidation purposes is distributed fairly to remaining landowners so they have more contiguous properties without roads going through them.

Different productivity classes of soil are taken into account before valuing land and making the consolidation and distribution. Total value of the new piece of land distributed to the landowner needs to match the value of the land taken from them. If the value of the land distributed to a farmer exceeds the value of the land taken, the farmer has to pay Norway's NPRA for the extra value. The reverse also is true.

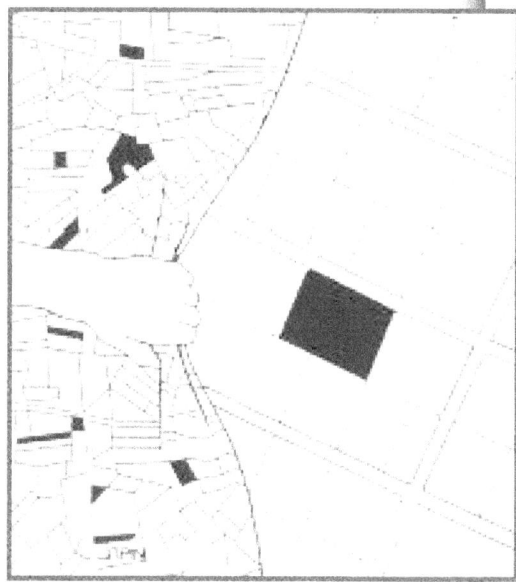

Figure 4. Land consolidation involves pooling and redistributing land.

Land consolidation involves pooling and distributing land, usually based on a zoning plan, to rearrange outdated or unsuitable layouts of land. In Norway, a special court of one professional and two lay people administer the program. One landowner can initiate an investigation, by means of a cost-benefit analysis, of whether land consolidation should be done. In Germany and the Netherlands, a majority of affected landowners is required to trigger such an investigation.

Acquisition of land by government agencies is partly for public infrastructure, but it also facilitates redistribution among private owners to achieve land use with a greater economic return. Land consolidation in Norway is commonly applied to agricultural land, but not urban land. In Germany, however, the team learned about a project in

Cologne involving considerable land consolidation that benefited both individual property owners and the city as a whole.

In the countries using land consolidation, the philosophy is that it is important to develop a land consolidation plan at the same time plans are developed for a road project. In Norway, farmland cannot be purchased without permission from public authorities. The government does not need a permit to buy a farm, but will generally pay 20 to 25 percent more than the market value of the property to carry out a land consolidation.

In general, landowners in the countries visited were pleased with the land consolidation concept. In most cases, they prefer to have land instead of monetary compensation, and land consolidation enables them to achieve farms with better layouts. Because of the way farmland has been divided over the years in Norway, a farmer may own several noncontiguous pieces of land. Land consolidation can concentrate a farm into one or two fields with no road through it.

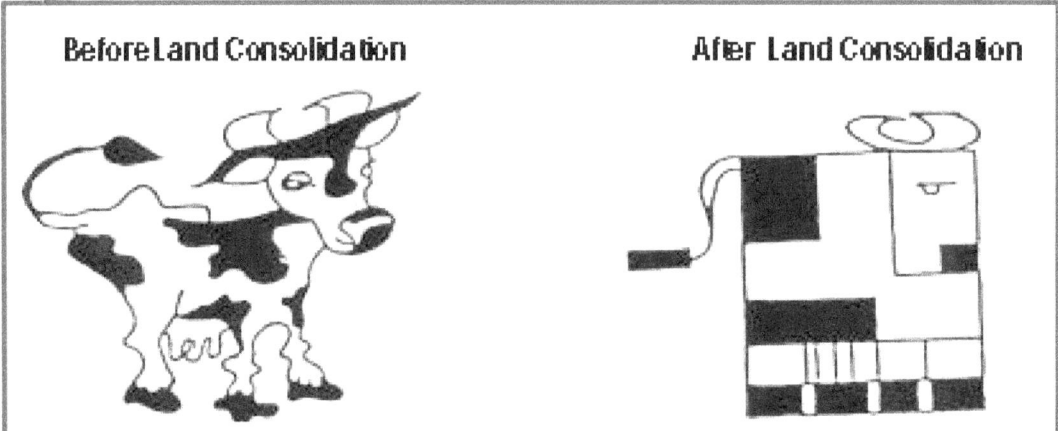

Figure 5. Land consolidation enables property owners to achieve farms with better layouts.

Chapter Four
TRAINING

Highway right-of-way staff positions are considered professional-level positions in all four countries visited. In Norway, all right-of-way personnel have a five-year degree specific to their profession. In Germany, most right-of-way staff members the team met with were economists, engineers, or attorneys. In England and the Netherlands, right-of-way personnel hold various college degrees and have opportunities to participate in strong continuing education programs. Training requirements vary, but all of the countries place a strong emphasis on formal training and continuous employee development.

PRIMARY FINDINGS

Norway's Five-Year Degree Program

Most of the top-ranking staff members in the NPRA's Land Acquisition and Real Estate Division are graduates of the Agricultural University of Norway. The university has 2,000 students in a five-year program that covers property rights and land law. In Norway, a student must complete 13 years of schooling before entering college, which means a university graduate has 18 years of formal education. The university program in land use planning dates back to 1898. The university has offered a curriculum leading to a career in right-of-way and land consolidation since 1960. It also offers master's degree and Ph.D. programs in land use planning.

Figure 6. Right-of-Way and Utilities scanning team members and their Norwegian hosts visited the Agricultural University of Norway in As.

Areas of study include landscape architecture, law, architecture, engineering, and land tenure and relocation. Land tenure and relocation involve the following subject matter:

- Land acquisition
- Eminent domain
- Land reallocation
- Land valuation and devaluation
- Negotiation
- Mitigation
- Related subjects

The team was impressed with the concept of a degree leading to a right-of-way career. Most of the curriculum for such a degree already exists at U.S. colleges offering degrees in real estate and civil engineering. It would be a small step to create a new combination of existing courses at these colleges and combine them with courses offered by the International Right-of-Way Association and the National Highway Institute. Identifying and offering the curriculum for a degree in real estate with a specialty in right-of-way should create interest among potential candidates who are now unaware of right-of-way career opportunities.

The Netherlands' In-House Training Programs

Universities in the Netherlands do not include a right-of-way degree in their regular curriculum. Universities are willing to provide tailor-made training, but the cost would be too high. More than 25 years ago, Rijkswaterstaat (Ministry of Transport, Public Works and Water Management) contracted with a training company to develop a training program for its right-of-way staff that leads to a certificate of completion of higher professional training.

The training company developed a two-year program of eight-to-10 one-day educational sessions every six months. First-year training starts with the legal system and covers the civil law code and other legal aspects affecting right-of-way acquisition. The second half of the year is devoted to general appraising and the types of appraisal problems confronted by Rijkswaterstaat. Second-year training covers right-of-way acquisition in depth, including the intricacies of complicated appraisals. After two years of study, students must pass an examination and write a paper.

Instructors are of three types:

- Right-of-way consultants who are experts in specialized areas of right-of-way acquisition
- Experienced appraisers who have done appraisal work for various types of government projects
- Instructors from the judiciary who teach courses in such areas as expropriation and use real-life examples from their own experience in their teaching

All right-of-way staff members are required to take the training program. Rijkswaterstaat employees wishing to work in the right-of-way field must apply to take the program, and agency officials select students they believe would be good at negotiating and dealing with other aspects of right-of-way work. It is universally recognized that maturity and people skills are necessary to make a good right-of-way agent. New employees are not eligible for right-of-way training until they have worked for Rijkswaterstaat long enough to demonstrate the aptitude and maturity required to be successful right-of-way agents.

Rijkswaterstaat pays for training for their employees, but the program also is open to individuals from other agencies and the private sector at their own expense.

Graduates of the program attend semi-annual meetings to discuss their experiences in the field and the impact of new developments on their work. These follow-up meetings help maintain the competency of right-of-way staff. Supervisors measure the effectiveness and competency of right-of-way staff in routine performance evaluations.

England's In-House Training Programs

The Highways Agency in England has an extensive training program for employees, including those in the Lands Branch. A central training team and a team of local training advisers has been established to administer:

- A central training program
- Tailor-made courses
- External training
- Support for external studies
- Learning resource centers with computer-based training

Each employee is authorized to receive two weeks of training per year. The Highways Agency supports external training. Although employees have to take it on their own time, the Highways Agency pays the fees.

The Highways Agency uses a performance appraisal system that provides annual reviews and is based on competencies needed to do the job. It establishes a framework for individuals and line managers to identify development and training needs, as well as provides interim reviews to ensure that development and training needs are kept in line with personal and business objectives.

The Highways Agency has a mentoring program to provide basic information on the agency to each new employee. The agency also has contracts with professional valuers, attorneys, and managing agents to provide training sessions for staff members. The agency issues regular updates and advice on policies and legislative changes, and arranges and finances additional specialty and job-related training. It promotes and disseminates best practices through seminars, which can be either activity specific or general. The Highways Agency uses professional development plans to determine the level of competency required in particular employment grades. Entry-level staff members normally have at least the minimum academic qualifications required for recruitment to the civil service.

OTHER OBSERVATIONS

German Association for Appraisers

The German Association for Appraisers is a large association of appraisers for agricultural properties. Many appraisers of improved and commercial properties belong to a Belgian appraisal association because there is no association of residential or commercial appraisers in Germany.

Royal Institution of Chartered Surveyors

England is the home base for the Royal Institution of Chartered Surveyors (RICS). In its booklet "World Class University Partnerships" is the following statement:

"Chartered surveyors are the leading source of professional advice on land, property, construction, and related environmental issues, on a global scale. Chartered surveyors are based in over 100 countries worldwide and are represented by national or regional associations in over 50 countries. They hold an internationally recognized qualification and belong to the premier property professional body in the world. And it is the profession's high standards of education and training which guarantee its universal value within the international business community. The Royal Institution of Chartered Surveyors is committed to developing strong partnerships with recognized centers of academic excellence throughout the world, to ensure the direction and growth of the profession in the long-term. Of particular interest are those centers which offer: excellent teaching faculties working closely with business, international standards of research, an appropriate range of courses at undergraduate and postgraduate level, and the ability to attract students of suitably high caliber."

RICS was founded in London in 1868 and is one of the largest professional associations in the world, with nearly 100,000 members, including 75,000 chartered surveyors (appraisers) and 25,000 undergraduates and graduates in the process of qualifying. RICS has accredited more than 60 courses in Europe, Asia, Australia, Africa, and the Caribbean. Key partners include the University of Cambridge, University of Melbourne, European Business School in Frankfurt and Berlin, University of Hong Kong, and University of the Sorbonne.

The RICS publication "The Official Prospectus to Surveying Education, Higher Education Courses in Land, Property and Construction" has a listing of undergraduate and postgraduate conversion courses accredited by RICS and the universities that offer them.

The RICS publication "Policy and Procedures for Accredited Centres and Accredited Courses" details its system of accreditation, which distinguishes between accredited centers and other centers offering accredited courses. The RICS publication "APC Requirements & Competencies" sets out standards that must be achieved to qualify for particular divisions of RICS or surveying specialties. APC stands for "assessment of professional competence."

Appraiser/Relocation Agent

In the countries visited, the appraiser is usually the relocation agent also. This was once a common practice in the United States, but over time those functions became separate. From a training perspective, this means that relocation agents in those countries are certified appraisers. Specialized training in relocation is not necessary, since relocation benefits in those countries are not set by regulation, but are simply whatever costs are necessary and reasonable for the resident or business owner to relocate.

Figure 7. Myron Frierson and Richard Moeller listen to interpreters at the German Federal Ministry of Transport.

Chapter Five
UTILITIES RELOCATION AND ACCOMMODATION

Scanning team members were impressed by utility practices they observed in the European countries, each of which uses strategies to better relocate and accommodate utilities located on or near highway rights-of-way. Most of the countries make special efforts to enhance relationships between highway and utilities officials by improving coordination, cooperation, and communication.

PRIMARY FINDINGS

While team members observed many worthwhile utility practices, they identified seven that are either new to the United States or not used here uniformly and have the greatest potential for improving results. These practices include cooperation, coordination, and communication; underground utilities; utility corridors; recognizing pipelines as a mode of transportation; avoiding unnecessary utility relocations; utilities in design-build contracts; and master utility agreements.

Cooperation, Coordination, Communication

Special efforts are being made in the countries visited to enhance working relationships between highway and utilities officials by improving cooperation, coordination, and communication.

Crowded conditions and time constraints in the Netherlands dictate the need for a cooperative, coordinated highway and utility environment. Although friction existed between highway and utilities representative in the Netherlands in the past, this is no longer the case. Over the past 10 years, the highway agency's utilities representative has forged a good working relationship with utility companies by emphasizing communication skills such as listening, asking questions, and discussing issues. The need for a teamwork approach to developing projects has been recognized not only between highway and utility personnel, but also within the highway agency. Proposed project teams include representatives from planning, design, environment, right-of-way, utilities, and construction offices.

In Norway, team members observed a high level of trust between highway representatives and utility owners. The team noted that the NPRA exhibited a helpful, caring, and—as one team member noted—almost maternal approach to working with property owners. This approach carried over to working with county administrators, consultants, contractors, utility companies, and others. As a result of this attitude, utilities are relocated without delaying highway contractors, disagreements over costs are few and easily resolved, and required oversight and inspections are minimal.

In England, the Highways Agency Codes of Practice—known as the Green Book—outline procedures to follow when utilities are affected by road projects. These procedures promote face-to-face meetings with utility representatives rather than letters or e-mail, and encourage give and take. Although Highways Agency and utilities personnel traditionally tended to stay aloof, they now recognize the critical need to work together and find common ground.

In addition, England has regional Highway Authorities and Utilities Committees (HAUC) that meet quarterly to discuss issues and upcoming projects that might affect planned utility work. Issues that cannot be resolved are taken to the national HAUC, which considers the best approach to resolving them. These discussions can get intense, but are considered valuable to parties involved.

In 1998, an FHWA survey determined that the most significant utility-related problem was lack of cooperation, coordination, and communication. Although some State departments of transportation work well with utility companies and some have active utilities coordinating committees, this is not true in all States. A report prepared by the U.S. General Accounting Office concluded that States with good cooperation, coordination, and communication had fewer utility-related problems.

European strategies in this area may be helpful to U.S. practitioners. The scanning team recommends that State DOTs continue and intensify efforts to emulate the attitudes and methods being used in Europe. This includes frequent face-to-face meetings with utility company representatives and a willingness to listen, ask questions, and discuss common problems to achieve an amiable working relationship.

Locating Utilities Underground

Utilities are placed underground routinely in most countries the team visited, especially on high-speed, high-volume highway facilities. The rationale for the practice is both safety and landscape aesthetics.

Except for high-voltage power lines, utilities on or along all highway rights-of-way in the Netherlands are installed underground. High-voltage lines are installed on large transmission towers located outside the highway right-of-way and well removed from the roadway. Power companies in the Netherlands have located their facilities underground for the past 40 years, making utility pole collisions virtually nonexistent.

Efforts are made in Germany to place utilities underground, especially in built-up areas, primarily to satisfy environmental requirements. In England, most utilities along the trunk roads—motorways and other principal arterial highways—are underground. New utilities are placed underground in Norway on the national and county roads, which are maintained by the NPRA.

Utility poles have become a significant safety problem in the United States, where an estimated 80 million poles are located along roads and streets. Every year, almost 1,400 motorists are killed and some 60,000 are injured in collisions with utility poles. A national task force has been established to investigate solutions to this problem.

Locating utilities underground in the United States would be costly and difficult in areas with rocky or unfavorable soil conditions and other constraints. Even so, European countries have proven it can be done and team members believe it deserves serious consideration in the United States. State DOTs should evaluate whether their decision-making process for utility installations looks beyond costs and gives appropriate weight to factors such as safety, environmental effects, and community aesthetics.

Utility Corridors

Several European countries are using or are considering establishing utility corridors for utilities desiring to cross highways or locate their facilities longitudinally along highway rights-of-way. Utility corridors consolidate utility locations, maximize use of limited available land, and minimize or eliminate road openings. Conduit can be placed in these corridors for future use by multiple utilities and joint trenching can be used to arrange multiple utilities in the same trench. In some countries, highway agencies design bridges to carry future fiber optics installations.

In Norway, utilities can locate their facilities on highway rights-of-way free of charge whenever space is available. On new highway construction projects, utility companies place conduits under the roadway for future crossings. Since open cuts and jacking under pavement are not allowed, all utilities crossing these highways must be placed in the conduits. Joint trenching is being considered for longitudinal installations, in which utilities desiring to use a right-of-way will coordinate arrangements to place all their facilities in one trench.

In the Netherlands, most utilities along roadways are on private land because normally only enough right-of-way is purchased to accommodate the highway cross section from ditchline to ditchline. An upcoming national plan will recognize utilities as a mode of transportation. Utility corridors will be established for utilities that want to cross motorways or locate facilities longitudinally on highway rights-of-way for short distances. Rijkswaterstaat will acquire the utility corridors and utilities will have to coordinate usage within them, but the question of whether utilities will have to pay to use them is unsettled. Another aspect of this effort is a review of safety standards and installation techniques to determine whether separation distances can be reduced to facilitate joint locations in utility corridors. Finally, the Netherlands is working to rationalize utility locations as projects necessitate relocations. One example of this is the Caland Tunnel Crossing. On this project, Rijkswaterstaat and utility companies developed joint locations that resulted in a more logical layout of utility installations, fewer river crossings, lower overall land requirements, and lower costs.

In the United States, highways and utilities sharing highway rights-of-way is considered to be in the public interest. Such use is subject to State DOT approval and the issuance of permits. All States prohibit cutting pavement to cross freeways and most prohibit cutting pavement to cross principal arterials. Some States have made efforts to locate utilities in designated utility corridors and encourage utilities to coordinate their work to better use available space. Most other States have tried to get utilities to place their longitudinal underground facilities as near the right-of-way line as possible in rural areas and behind sidewalks in urban areas. Utilities installed longitudinally on highway rights-of-way in most States, however, are still located in other places, even under the pavement and shoulders on some facilities.

As more utilities in the United States desire to cross or use highway rights-of-way, it will become imperative for State DOTs to manage the use of the right-of-way more effectively. This may involve establishing utility corridors and requiring utility companies to coordinate installation of facilities within these corridors.

Recognizing Pipelines as a Transportation Mode

A new national plan in the Netherlands recognizes pipelines as a mode of transportation, along with highways, air, marine, and rail. This will have many implications, starting with greater use of gas transport lines followed by increased use of water, electrical, mineral, and petroleum transport lines.

While the focus of the new Dutch policy is gas pipelines, the concept may have broad applicability to other products. Many types of material now transported by freight truck could be moved through pipelines as liquids or slurries, reducing traffic on roadways.

As highways in the United States become more congested and air quality concerns increase, using pipelines instead of trucks to transport essential products may be beneficial. The role of highway agencies in fostering this new mode could include facilitating research and development of methods to exploit this type of transport, establishing routes and corridors for pipeline companies to use, and funding construction and operation of pipelines.

Avoiding Unnecessary Utility Relocations

The German Lander, or states, manage national trunk roads on behalf of the Federal Ministry of Transport. This includes planning, preliminary engineering, right-of-way acquisition, utility relocation, construction, and maintenance. The Lander make every effort to avoid relocating utilities during highway construction by identifying all utilities early in project development and designing around them where possible.

U.S. projects traditionally have been designed with little regard for utility locations. If utilities were in the way of a new or reconstructed highway, they had to move. This costly practice is being replaced slowly because of growing awareness that unnecessary utility relocations are not in the public's best interest. Many State DOTs have avoided unnecessary utility relocations with an engineering practice called subsurface utility engineering, a process through which they obtain comprehensive underground utility information. With this information, highway designers can make minor design changes to avoid many underground utilities. A recent Purdue University study revealed that State DOTs saved at least $4.62 in avoided costs for every $1 spent for subsurface utility engineering. Savings to utility companies and the public are believed to be even greater. The scanning team recommends that State DOTs explore this and further efforts to avoid unnecessary utility relocations.

Utilities in Design-Build Contracts

England's Highways Agency plans to advertise 19 design-build contracts for major projects over the next three years. Utility relocation is an essential part of these projects, although the Highways Agency's contractor rarely does utility installations because of qualification issues. Design-build contracts are advantageous to the Highways Agency because risks of utility-related delays are transferred to highway contractors, reducing claims for delays and the cost of large overruns. Highways Agency personnel stress that good estimates of utility costs are essential because the agency must pay the contractor estimated costs no matter what actual costs turn out to be, even if they are lower than the estimate. Overruns up to 20 percent over the estimate are the contractor's responsibility, and the contractor and the Highways Agency share anything beyond that. Utility companies receive the actual cost of the work performed.

Under the design-build concept, the contracting agency identifies parameters and establishes design criteria for a project. By allowing the contractor to optimize workforce, equipment, and scheduling, the design-build concept allows for greater flexibility and innovation. Along with increased flexibility, the contractor must assume greater responsibility. From the contracting agency's perspective, the potential timesaving is a significant benefit. Since design and construction are accomplished through one procurement, construction can begin before design details are finalized. Also, because both design and construction are performed under the same contract, claims for design errors or construction delays because of redesign are not allowed and the potential for other types of claims is reduced.

In the United States, about 21 State DOTs and several local transportation agencies have design-build projects approved or under way. Some projects include utility relocation activities. Highway contractors may find utility relocations in design-build contracts difficult at first, because they have not had an opportunity to establish credibility with utilities and they are inexperienced in coordinating utility activities and designing and installing utility facilities. Even so, these problems can be overcome with experience, and the potential for saving time by reducing delays is enormous. The scanning team recommends that utility relocations be included in design-build contracts whenever possible.

Master Utility Agreements

Master utility agreements between the highway authority and utility companies are used commonly in Germany. These agreements—which outline authority, obligations, and liabilities—are used in lieu of individual project agreements.

The AASHTO Subcommittee on Right-of-Way and Utilities has established master agreements as a best practice because they eliminate the need to obtain approvals on every contract and save time for DOTs and utility companies. The scanning team recommends that State DOTs not already doing so seriously consider using master agreements. AASHTO and FHWA should consider developing model master agreements or distributing sample master agreements from State DOTs now using them.

OTHER OBSERVATIONS

Utility Installations by Highway Contractors

In Norway and England, highway contractors sometimes place conduit for utility companies. This also occurs on some projects in the United States. The AASHTO Subcommittee on Right-of-Way and Utilities has established utility installations by highway contractors as a best practice.

Highway contractors or their subcontractors can readily perform utility work, such as laying conduit for later use by utilities, installing storm and sanitary sewers, and laying water lines. While power, communications, and high-pressure pipeline companies may be reluctant to delegate work on their facilities because of safety, union, proprietary, or other concerns, they may be willing to allow the use of pre-approved subcontractors. Highway contractors may likewise be reluctant to assume responsibility for work outside their normal qualifications or experience.

Utility installations by highway contractors allow contractors to better control work and coordinate sequential or concurrent operations, reducing delays and disruptions. This practice also allows greater use of the highway contractor's equipment and manpower, reduces duplication of effort on items such as traffic control, and lowers bid prices by consolidating items such as excavation in one contract.

CHAPTER 5

Cost Sharing

Although most utilities in Europe are privately owned, they normally are allowed to occupy public right-of-way if they serve the public. Except in England, if utilities need to relocate to accommodate highway construction, they usually must do it at their own expense.

When utilities on highway rights-of-way in England must relocate to accommodate construction, utility companies pay 18 percent of the relocation cost and the Highways Agency pays the remaining 82 percent. The Highways Agency also pays 82 percent of the cost of preparing estimates, preliminary engineering, advance materials orders, supervision, inspection, overhead, and other eligible expenses. Utility companies normally pay for these activities initially, sometimes before a project is approved, and are later reimbursed by the Highways Agency. If desired, 75 percent of the Highways Agency's 82 percent may be paid to the utility company in advance, either by lump sum or installment.

Germany uses a different approach. When a utility has a property interest in its present location, the German highway agency is obligated to pay for the relocation. When the highway agency and a utility disagree on relocation costs and the dispute threatens to delay work, the highway agency will advance construction costs to the utility under a pre-financing agreement. Once the compensation question is settled or determined by a court, the utility returns any overpayment to the highway agency.

In the United States, when utilities located on public right-of-way are required to relocate to accommodate highway construction, the utility companies normally pay 100 percent of relocation costs. Exceptions to this are in New Jersey and Alaska, where the State DOT pays 100 percent of relocation costs, and in Montana, where the State DOT pays 75 percent of relocation costs.

England's 18/82 and Montana's 25/75 cost-sharing arrangements have potential advantages. When the highway agency pays most of the cost, utilities are more likely to be relocated in a timely manner. Another issue is that in the United States it is often difficult to determine who has prior rights, the highway or the utility. Often there are no good records to show whether the utility company has a property interest in its location. A cost-sharing arrangement covering every utility facility, regardless of location, could be useful in advancing projects to construction and avoiding unnecessary disagreements. In addition, State DOTs should consider paying preliminary engineering costs up front in exchange for an agreement from utilities to complete work in a timely manner. They also should consider paying construction costs up front in cases where there are disagreements over costs and the project is in danger of being delayed.

Right-of-Way Acquisition for Utilities

Several countries the team visited acquire right-of-way for utility purposes. The NPRA's Land Acquisition and Real Estate Division acquires all right-of-way shown on Norwegian project plans, including right-of-way for utilities when NPRA designers determine it is in the best interest of the project. In Germany, the highway agency often buys extra right-of-way for utilities. Right-of-way cannot be purchased solely for utilities in England, but sometimes replacement right-of-way is included in negotiations when the utility company has a property right.

When State DOTs acquire right-of-way in the United States, FHWA encourages them to consider consulting with affected utilities and acquiring sufficient right-of-way to accommodate utility needs. In addition, AASHTO's Subcommittee on Right-of-Way and Utilities considers the acquisition of right-of-way for utilities purposes to be a best practice.

When they intend to permit utilities to occupy highway right-of-way, some State DOTs consider this use in determining the extent of right-of-way needed for the project and acquire additional right-of-way solely for utility purposes. They may keep the acquired right-of-way, or sell, lease, or otherwise convey it to utilities. The scanning team recommends that State DOTs that do not acquire right-of-way for utilities seriously consider doing so. This practice will minimize inconvenience to property owners created when both DOT and utility representatives approach them to acquire property rights for a project.

Damage Prevention

Excavation activity causing damage to underground utilities is a problem in Europe. Utility companies in Germany are responsible for identifying their underground facilities and providing this information to highway contractors before excavation. If utility companies cannot provide this information, they must physically uncover facilities at their own expense to obtain the information. Germany is considering adopting a nationwide uniform documentation system for all utilities.

Efforts are also made in the Netherlands to avoid damaging underground utilities during excavation. Highway contractors are required to call the Cable Tube Information Center before they begin excavation activities, and the center provides names of utility companies with facilities in the excavation area. The contractor must then contact each utility company, which must provide information on the location of their facilities.

Highway contractors in England also must notify all affected utilities before they begin to excavate. England is creating a computerized registry of all utility installations. Despite these efforts, damage to underground utilities continues to occur.

Extensive programs have been developed in the United States to reduce damage to underground utilities caused by excavation activities. One-call notification centers have been established in every State. Contractors are required by law to call these centers and provide appropriate information before they excavate. The centers must then notify all affected utilities. When notified, utilities must visit the proposed excavation site within a specified time and mark the location of their facilities with paint or flags. After calling notification centers, contractors must wait for the site to be marked, protect markings after they are placed, and hand dig within a specified distance—about half a meter—on either side of marked lines.

AASHTO's Subcommittee on Right-of-Way and Utilities considers participation in one-call notification programs to be a best practice. To protect underground utilities from damage, the scanning team recommends that State DOTs use one-call notification centers at an appropriate level of participation and provide oversight to assure that contractors participate fully in one-call notification programs.

Protected Highway Designation

In England, the Highways Agency has the power to designate a roadway as protected, precluding new utility installations. This designation applies to all motorways (equivalent to the U.S. Interstate System) and some trunk roads (roughly equivalent to the U.S. National Highway System). Where a road with utility installations is designated as protected, utilities may repair existing facilities and make service connections, but may not expand or replace facilities without moving outside the right-of-way. This protection is primarily for safety and operational efficiency.

As DOTs in the United States search for ways to increase highway capacity and facilitate traffic flow on the National Highway System, they should consider ways in which this idea may apply. For example, a tiered system of protection geared to highway classification may provide a tool for reducing traffic conflicts and safety hazards on the most important roadways in each State.

Minimizing Pavement Cuts

Pavement cuts are a significant problem in Europe. Motorways and most other trunk roads are protected from road openings made by utilities, but often lower-volume roads and streets are not. In Germany, underground utility crossings on trunk roads are made by boring, jacking, directional drilling, or similar means. Pavements are cut for utility crossings on lesser roads and to accommodate fiber optics on streets, a trend that is becoming a concern to road officials.

Pavement cuts also are a problem in the United States. Pavements on low-volume, low-speed highways are cut for utility crossings routinely in rural areas. In cities, pavements frequently are cut to access the many utilities located longitudinally beneath streets. Fiber optics installations are becoming particularly troublesome as streets are torn up for installations and then poorly repaired. Even excellent repairs are not sufficient to prevent reduced pavement life in cut pavements.

The scanning team recommends that the United States make a greater effort to use trenchless technologies for highway and street crossings, and to control the frequency of pavement cuts to access or install utilities under city streets.

Geographic Information Systems

Geographic Information Systems (GIS) are being used in Europe for mapping right-of-way properties. In Norway and England, software programs have been developed and GIS is being used extensively for right-of-way mapping.

Similar mapping activities are under way in the United States, including some efforts to use GIS for utilities. More research into the feasibility of using GIS to map utilities needs to be initiated.

Accommodation of Fiber Optics and Wireless Telecommunications

Fiber optics and wireless telecommunications are used extensively in Europe. In each of the countries visited, special legislation governs telecommunications installations on highway rights-of-way.

Norway's policy prohibits fiber optics or wireless telecommunications towers from being placed on highway rights-of-way, although this policy may be revised. A proposed fiber optic installation on highway right-of-way is being considered from the border with Sweden to Oslo, and so far the NPRA has made no plans to require cash or services in exchange for allowing this installation.

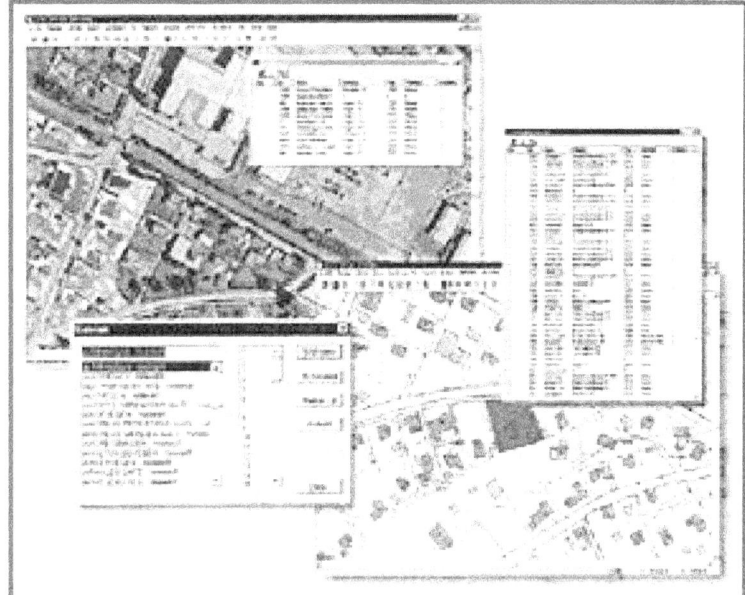

Figure 3. The Norwegian highway agency uses GIS software for right-of-way mapping.

Germany's Telecommunications Act includes provisions for installations on highway rights-of-way. Telecommunications lines, including fiber optics, may be installed free of charge along the right-of-way of the autobahn and other trunk roads. Wireless telecommunications towers may also be installed on the right-of-way, but they are not considered public services and charges can be levied. Neither fiber optics nor wireless telecommunications towers have been installed yet in Germany. Cables used by mobile phone operators have been installed in some tunnels for a fee, and wireless telecommunications towers have been installed in some of the privately owned service areas along motorways.

Procedures are being developed in the Netherlands to accommodate fiber optics and wireless telecommunications towers. Compensation has not been seriously considered, but provisions will be made for this in the new procedures.

Consideration is being given in England to accommodating fiber optics and wireless telecommunications towers in exchange for cash or services.

Chapter Six
Project Development

In the four European countries visited, typical time frames for project development are longer than in the United States. True comparisons are difficult, though, because project development in the European countries includes the planning phase. While the emphasis in the United States remains on reducing the time required for project development, the scanning team found a number of practices States could adopt or modify to promote improved project development practices and results.

PRIMARY FINDINGS

Multidisciplinary Team Approach

Several countries employ a project management approach, including the use of multidisciplinary project teams. Teams in the Netherlands are responsible for a project from planning through construction. Other Dutch management practices include:

- Right-of-way and utilities participation beginning at the planning stage.
- Budget and schedule commitments with a sign-off by functional representatives and project managers.
- Treating right-of-way activities as a critical path element of project management.
- Accountability for delivery on commitments.

In some countries, such as England, the project team is in a separate part of the transportation agency. The British use a framework document to facilitate coordination and communication by defining the respective roles and responsibilities of lands acquisition personnel and project team members.

Benefits cited as a result of using a project management approach include a shift in employee loyalty from functional units to the project as a whole, better communication and coordination among disciplines, more realistic scheduling, and earlier problem identification and solution.

Design-Build

England uses design-build contracting extensively in its program to reduce the time required for project development. While contracts include utilities coordination, right-of-way acquisition remains with the Highways Agency. Germany established an agency (the German Unity Planning and Construction Company for Trunk Roads, known as DEGES) to expedite new construction and rehabilitation projects important to the reunification of the country. The responsibilities of DEGES include land acquisition on behalf of the transport agency.

These practices suggest transportation agencies in the United States may benefit from considering further innovations in the area of design-build contracting, including expanding the design-build contract scope to include some or all right-of-way services.

Multidimensional and Inclusive Planning Processes

In several countries, zoning and land use plans prepared at the local or regional level govern decisions about the location of transportation infrastructure. Transportation agencies normally have a role in developing the plans. Redevelopment and transportation infrastructure issues are considered at the same time. Land use planning and modal integration are major focal points for the transportation agencies and others. The processes in each nation provide for significant input from affected property owners, community members, and local authorities. The planning efforts lead to adoption of a detailed definition of the project before involuntary right-of-way acquisitions begin.

Norway's planning system operates at county, municipal, and zone levels under its Planning and Building Act. The zone plan is the most detailed and complete. Road decisions normally are made based on municipal, and sometimes county, planning. The planning process includes participation from an array of interested parties, including landowners.

Germany uses a plan settlement and approval process when projects may significantly affect private parties and there is opposition to the proposed infrastructure. The process includes a public hearing before an independent authority that balances public and private interests, including the needs of utilities. The specific procedures depend on the scope of the project. Two simplified processes are used when the transportation proposal itself is insignificant or there are no significant impacts or opposition, property owners have agreed to the necessary acquisitions, and consensus has been reached with others on matters of public concern. An approved plan sets the alignment and right-of-way for the roads.

The Tracewet (Route) Act has defined the decision-making process for projects of national importance in the Netherlands since 1994. Under this act, Rijkswaterstaat assesses and balances economic and environmental issues, includes significant public participation in plan development, and looks for agreement with local authorities on proposed projects. Projects must be consistent with local zoning plans, although such plans often are revised to achieve needed consistency. Amendments to the Route Act in 2000 are expected to permit the route plan to prevail over the local land use plan when local agreement and land use plan revisions are not achieved within a reasonable time. A benefit of the zoning plan requirement is that it prevents inconsistent land development from occurring while zone plan changes are pending and after revisions are adopted. It also fosters public acceptance of proposed projects.

In England, more location and design detail is added at each level of planning, with county-level plans being the most explicit. At that level, the plan typically identifies access points along major highways, or trunk roads. The Highways Agency has the power to direct the regional planning authority to deny or grant access with conditions. No general legal right of access to trunk roads exists in England.

A major benefit of strong local planning systems in the countries visited is the broad ability to make thoughtful and comprehensive decisions about future needs, including appropriate land use and transportation infrastructure. The system also improves project quality and public support, and creates the opportunity to save considerable time in the project development process. The success of European practices suggest that re-examination of corridor preservation is warranted in the United States, using the "1990 Report of the AASHTO Task Force on Corridor Preservation" as a starting point. The review should consider how States might benefit from lessons offered by the more holistic European approach to land use, environmental, and transportation planning.

Definition of Problems and Solutions

To improve project quality and save time, European nations typically use the planning process to define specifically problems a project will address and how it will achieve intended results. The objective of this practice is to prevent scope creep, unnecessary work, and late plan changes. The scanning team found the most rigorous example of this in the Netherlands, where the order for property expropriation includes a description of identified transportation problems and a statement specifying how the project will address those problems.

Planning Stage Feasibility Analysis

Several countries use broad feasibility reviews before right-of-way acquisition. Items considered in the Netherlands include land use, environmental effects, financing, and engineering. Germany does a similar review that incorporates a cost-benefit analysis of traffic and safety measures.

Land Consolidation

A European concept that caught the attention of the scanning team is the land consolidation process. Norway, Germany, and the Netherlands all use some form of this practice. Land consolidation allows pooling of fragmented land parcels and redistribution using more economically rational parcel configurations. Distribution is based on value, so owners receive land of the same value as the land they put into the pool. This procedure is used primarily in agricultural areas to reorganize properties where a project has an adverse effect because of a new alignment or significant widening. Germany also uses the process to implement a new urban zoning plan and its accompanying transportation infrastructure. In Norway, one owner's request can initiate the process, while in Germany and the Netherlands a majority of affected owners must agree to start consolidation.

Although this practice may seem foreign to Americans, European property owners are pleased to have more economic parcels and new roads. The highway agency benefits because it can reduce the number of highway crossings needed to service parcels separated by the road. Used on a voluntary basis, the land consolidation concept may offer promising opportunities in the United States to improve land use and property operating characteristics after a highway project is completed.

Realistic Right-of-Way Budgets and Schedules

Each of the European countries studied typically allocates enough time and money to permit appropriately timed and scoped acquisitions. Each country has its own version of the British principle: put individuals affected by a project in positions in which they neither gain nor lose from the project. This operating rule leads to an owner-oriented process, including broader use of flexible acquisition benefits and property management practices. One example is the British approach under which strip acquisitions with property effects unacceptable to the owner can be mitigated through full acquisition and resale of the affected property.

In all four countries, a high priority is placed on settlement of acquisition cases. A premium often is paid to accomplish settlements. A Rijkswaterstaat representative in the Netherlands referred to this practice as making an investment decision that takes into consideration the various costs and benefits of expediting a settlement.

Overall, the European approach results in better processes and outcomes for property owners than often occur in the United States. Property owners find themselves in at least as good a position as they were before the project and sometimes in a better situation. Settlement rates and abutter satisfaction rates are high, which helps to avoid project delays.

External Communication, Coordination, and Participation

Each country visited engages in extensive public coordination, and they each characterize the practice as valuable. The countries consistently encourage owner participation in design issues at early stages of project development. Several countries typically have the project manager or designer and the right-of-way team member conduct field reviews with affected property owners early in project development.

In England, designers are available to meet with property owners to discuss issues and impacts once the Highways Agency determines the preferred route. Before construction begins, the contractor holds a public meeting to discuss the work and its expected impacts.

Project teams in the Netherlands contact potentially affected owners early so their concerns can be addressed as the design is developed. The Netherlands is piloting a decision-making process for use on major projects called interactive planning. The goals of the interactive planning process are to speed up projects, gain more public support for projects, and find more creative transportation solutions through participative problem-solving sessions. Results suggest that a range of public participation models—including information, reaction, research, consulting, participation, and partnership—is needed to address the variety of circumstances affecting project development and implementation.

Broad public participation benefits the transportation agencies by helping to identify issues and incorporate needs and solutions into the original project design. This practice avoids many late plan changes and improves relationships with affected property owners, municipalities, and other parties.

Flexible Early Acquisition Tools

All four countries the scanning team visited have broad early acquisition authority. These early acquisitions provide the Europeans with great flexibility, saving time and money and reducing conflicts with affected property owners.

The British system permits voluntary acquisitions before a preferred route is announced. Once the preferred route is published, the system requires the Highways Agency to protect the route through discretionary and blight purchases. The British government acknowledges the potential for bureaucratic-commitment effects on the project decision-making process. It considers them minimal, however, and believes it is more effective to do early acquisitions knowing that the Highways Agency may have to sell unneeded properties if a project is canceled or relocated.

England also uses a compulsory purchase, or expropriation, order to open a process under which the Highways Agency can notify owners of the commencement of negotiations, or notice to treat. That notice triggers a right of compensation for owners, including up to a 90 percent payment based on an advance valuation estimate, and a three-year right of entry for the Highways Agency. The title transfers when the Highways Agency makes final payment. This allows the project to move forward without waiting for the resolution of land acquisition disputes.

In Norway, NPRA uses construction permits to authorize entry to do work while the acquisition process is pending. A negotiated advance payment accompanies the permit. Germany permits the voluntary purchase of land at any time, although this usually occurs after the planning process ends. Like Norway, Germany frequently uses a construction permit process to gain possession before title and compensation are final. The Germans have a process that can compel transfer of possession, but it applies only where there is a strong need for immediate use of the property. In both cases, the owner receives estimated compensation at the time of the transfer of possession. Rijkswaterstaat in the Netherlands can do voluntary acquisitions at any time. Expropriations can begin after the first decision of the minister to approve the route plan and do the project.

Another early acquisition tool used in Norway is a strategic acquisition fund that facilitates purchases when project budgets or acquisition schedules are not yet firm. Some U.S. States have tried this method. Like Norway, they found the process useful but limited by a chronic inability to maintain funding.

Use of early acquisition practices in the United States similar to those observed in Europe could save significant project development time. The scanning team proposes that State highway agencies support consideration of amendments to Federal laws and regulations to authorize broader use of the techniques. The objective is to develop a system of early acquisition that protects the integrity of the project decision-making process, yet permits projects with limited right-of-way impacts to move forward as categorical exclusions or with some other type of qualified exemption from National Environmental Policy Act constraints on pre-approval acquisitions.

User-Friendly Right-of-Way Plans

Right-of-way plans observed in Europe are clearer and easier to interpret than plans in many jurisdictions in the United States. In Norway, for example, zone plans are the basis for acquisition. Norway also does minor acquisitions without plans. The Netherlands negotiates with property owners using a general schematic showing existing and proposed right-of-way lines and macro-level design information. Parcel-specific acquisition information is shown on a land interest plan that excludes engineering information. The Netherlands does not require final design to move forward with voluntary acquisitions, but detailed plans are used when involuntary acquisitions are done by expropriation.

In England, an engineering schematic, land interest plan, and occasionally a model are used at public hearings. The land interest plan shows only the area to be acquired, field reference numbers, and boundaries of the acquired properties. It is the plan used for property owner negotiations and for recording at the registry of deeds. No longitudinal baseline is used on the plans and no reference points are plotted in the field. A unique GIS center point is used for the parcels acquired. The Highways Agency uses as-built plans of the roadway to locate points in the field after a project is completed. Detailed construction or survey plans may be used in litigation if needed.

Countries using land interest plans report that plan simplicity is not only acceptable to owners, but makes it easy to explain acquisitions to them. Negotiators typically use engineering schematics to explain construction impacts.

One important element that facilitates use of these simpler plans is the national standardized mapping, land registration, and survey system in each country. Nonetheless, European practices suggest that there are opportunities for cost savings and simplification that could be exploited by those States now using highly detailed and complex right-of-way plans for negotiations and acquisition documentation.

Right-of-Way Databases and GIS Systems

Several countries have projects under way to develop systems for managing data relevant to right-of-way functions. Norway has an integrated GIS system for right-of-way work that includes property data by parcel, environmental data, property maps, and geo-referenced images of corrected vertical photographs. The system can combine data types and produce three-dimensional images. Norway makes the system available for public use at the highway agency and the software is commercially available.

England is developing a database system called the Terrier that will include deed descriptions, GIS data, and government survey grid information for each of the approximately 100,000 parcels the Highways Agency owns. England also has a property management database for all properties and is developing an Acquisitions Information and Estimates System that will tie into the Terrier.

The scanning team found such systems useful and recommends that State DOTs explore opportunities for developing a system for their use.

OTHER OBSERVATIONS

The scanning team observed several other project development practices that merit attention.

Norway has standard right-of-way acquisition limits. Its minimums are one meter from the backslope of the ditch and three meters from the edge of pavement. This simplifies the design and acquisition processes.

England handles construction-period mitigation and improvement work for property owners as a form of compensation. This accommodation work is treated as an element of payment whether the work is done directly by the Highways Agency contractor or by the owner's contractor. The limit of accommodation expenditures is the value of damage to the property. This formalization helps avoid double-payment cases in which owners are compensated during acquisition for damage the agency's construction forces subsequently cure.

CHAPTER 6

Germany's land consolidation system provides for conveyance of the right-of-way for roads to the municipality for free, in recognition of the enhanced property values created by the future road system.

The Netherlands sometimes uses special legislation to create fast-track planning and approval processes for projects of high national priority.

Figure 9. Joachim Pestinger (left) meets with German delegation members Peter Weitershagen, Hans Mundry, and Hans Stumpel.

Chapter Seven
RECOMMENDATIONS

The Right-of-Way and Utilities scanning team discovered many similarities between the right-of-way and utilities practices in the European countries they visited and those in the United States. The team also observed a number of interesting and innovative practices in those countries that, if implemented in the United States, could benefit State and Federal transportation right-of-way and utilities programs.

The team met on the last day of the scanning study to review their findings and identify practices with the greatest potential for implementation in the United States. Their recommendations and implementation strategies are in this chapter. Implementation on some strategies had begun already when this report was published. Chapter Eight details those activities. The findings, observations, conclusions, and recommendations are those of the scanning team and not FHWA.

APPRAISAL AND ACQUISITION

Early Involvement of Property Owners in Design Process

Early and frequent contact with property owners is essential to successful acquisition. The scanning team encourages States to consult affected property owners before completion of project design to assess the impact of the proposed design and to determine if a design revision is warranted. Selective use of this practice could result in more timely purchases and reduce damages to the affected properties.

Property Owner Interviews

The team encourages acquisition staff to use an in-depth interview process, when appropriate, to discuss the impact of a project with property owners. This interview will afford a better understanding of how owners use the property. Information obtained from the interview can be used to determine if further investigation into possible damages is necessary. If so, appropriate experts should be assigned to assess the project's impact on the property. Findings from the in-depth interview, appraisal, and expert analysis will help form a comprehensive estimate of just compensation, which will facilitate negotiations with property owners.

Limited Use of Appraisal Reviews

States are encouraged to adopt a risk management-based appraisal review system similar to those used in some European countries. The goal is to determine whether such a system—for example, auditing a sample, reviewing all complex appraisals, or setting review thresholds—can protect the quality and integrity of the valuation process while saving overall project time and costs. The team recommends that a risk management-based appraisal review system pilot be undertaken in several States in conjunction with FHWA. Results of the pilot should be used as a basis for any regulatory changes.

Appraisal and Negotiation Functions Performed by Same Person

In Europe, the same person commonly performs both appraisal and negotiation functions on a parcel. The team recommends that a pilot program be conducted in several States in which the same individual conducts both appraisal and negotiation functions when acquiring land for right-of-way. The goal of the pilot is to determine if such an approach is cost effective while at the same time assures appropriate treatment of property owners.

Incentive Payments

Based on the European approach to negotiations, the team recommends that emphasis be placed on compromising on issues related to just compensation. It is recognized that such techniques will effectively resolve acquisitions in a timely and cost-effective manner.

COMPENSATION AND RELOCATION

Voluntary Land Consolidation Pilot Program

Three of the four countries the team visited use some form of land consolidation. This process allows pooling of fragmented land parcels and redistribution using more economically rational parcel configurations. Owners receive land of at least the same value as the land they put into the pool. The team recommends that FHWA research the ability of States to accomplish voluntary land consolidation and implement a pilot program to evaluate the benefits.

Business Reestablishment and Relocation

The team recommends evaluating items eligible for business reestablishment and relocation reimbursement in the Netherlands and England. The European experience and results of the recent FHWA-sponsored research on "Business Payments and Services" can be used to support changes in Federal legislation and regulations. This could involve consideration of:

- Liquidation of small businesses beyond the $20,000 fixed payment now available under the Federal Uniform Act.
- Giving special consideration to business owners of retirement age.

Residential Relocations

The Norwegian model for compensation calculates a total payment based on replacement cost. This amount includes the estimated fair market value and any additional amount needed to replace the acquired property. This one-step process would save time and cost, and could be tried experimentally in the United States.

TRAINING

Pre-Employment and Employee Education and Training

Although training requirements vary in the countries visited, they all emphasize formal training and ongoing employee development. The team recommends that FHWA encourage establishment of a pre-employment degree program and employee education and training programs. This involves exploring the potential for recruiting one or more colleges to provide this service, including a college degree program for right-of-way careers and a continuing education program using distance-learning techniques. This proposal expands on the Federal Government's potential establishment of a real estate services academy.

The team recommends establishing a panel of representatives from FHWA, IRWA, AASHTO, and a private consultant to pursue this training concept. FHWA will act as the lead to contact colleges and online learning centers, with the goal of developing and implementing such a curriculum by Fall 2002.

Mentoring Methods

The team recommends evaluating mentoring activities in each State through AASHTO's Internet web site, developing a summary of mentoring methods in the United States and Europe, and recommending adoption by States.

UTILITIES

Cooperation, Coordination, and Communication

The team noted special efforts in European countries to enhance cooperation, coordination, and communication with utility companies. In the United States, according to the U.S. General Accounting Office, States with active utilities coordinating committees that meet regularly to discuss common problems have fewer utility-related problems than other States. The team recommends that State DOTs continue and intensify efforts to meet with utility company representatives regularly. DOTs should take the lead in developing and supporting utilities coordinating committees.

Underground Utilities

Utilities in most of the countries visited are routinely placed underground. As a result, utility pole collisions are not a problem. In the United States, utilities are located underground for aesthetic rather than safety reasons. Some State DOTs, however, are addressing safety by developing and implementing utility pole safety programs designed to systematically relocate hazardous utility poles.

The team recommends that State DOTs continue to develop or enhance utility pole safety programs. Locating utilities underground should be considered as a possible countermeasure, although putting all utilities underground would be costly and difficult in some States because of unfavorable soil conditions. State DOTs should evaluate their decision-making process for utility installations, looking beyond construction costs to give appropriate weight to factors such as safety, environmental effects, and community aesthetics.

Utility Corridors

The team noted that several European countries are using or considering establishing utility corridors for utilities crossing major highways or located longitudinally along highway rights-of-way. Conduit can be placed within these corridors for future use by multiple utilities, or joint trenching can be used to systematically arrange multiple utilities in the same trench.

In the United States, it is considered in the public interest for highways and utilities to share highway rights-of-way. Such use is subject to State DOT approval and the issuance of permits. As more utilities desire to cross or use highway rights-of- way, State DOTs should consider establishing utility corridors and requiring utility companies to coordinate the installation of their facilities within these corridors.

Recognizing Pipelines as a Transportation Mode

In the Netherlands, a new national plan recognizes utilities as a mode of transportation. As such, more credence will be given to the contribution of utilities to transportation. Initially this will include more use of gas transport lines, followed by increased use of water, electrical, mineral, and petroleum transport lines.

As highways in the United States become more congested and air quality concerns increase, taking advantage of pipelines to transport essential products normally transported by trucks may be beneficial. The team recommends that State DOTs consider methods to foster pipelines as a transportation mode. This could involve facilitating research and developing methods to exploit pipeline transport, including establishing routes and corridors for pipeline companies or funding construction and operation of pipelines.

Utilities in Design-Build Contracts

In the three years following the scanning study, England's Highways Agency planned to advertise 19 design-build contracts for major projects. Utility relocation is an essential part of these projects.

Contracts for design-build projects in the United States often fail to include utility relocations. By including utilities in these contracts, risks of utility-related delays would be transferred to highway contractors, reducing delays and large cost overruns. The scanning team recommends that State DOTs not already including utilities in design-build contracts consider doing so.

Master Utility Agreements

Master utility agreements between the highway authority and utility companies are used commonly in Germany in lieu of individual project agreements. These agreements outline authority, obligations, and liabilities.

The AASHTO Subcommittee on Right-of-Way and Utilities has established master agreements as a best practice because they eliminate the need for approvals on individual contracts. These agreements save time for both DOTs and utility companies, including the time necessary to consummate agreements.

The team recommends that State DOTs not using master utility agreements consider doing so. AASHTO and/or FHWA should consider developing model master agreements or distributing sample master agreements obtained from State DOTs that use them.

Additional Recommendations for Utilities

In addition to the primary recommendations described above, team members identified other utilities strategies that may have value in the United States:

Utility Installations by Highway Contractors—In Norway and England, highway contractors sometimes place conduit for utility companies. This also occurs on some projects in the United States. State DOTs, in conjunction with utility companies, should consider allowing highway contractors or their subcontractors to install such items as conduit for later use by utilities, storm and sanitary sewers, water lines, and possibly power, communications, and high-pressure pipelines. This will improve highway contractors' ability to control workflow and coordinate sequential or concurrent operations, reducing delays and disruptions.

Cost Sharing—In England, when utilities located on highway rights-of-way are required to relocate to accommodate highway construction, utility companies and the Highways Agency pay 18 percent and 82 percent of relocation costs, respectively. This cost includes preparation of estimates, preliminary engineering, advance materials orders, supervision, inspection, overhead, and other eligible expenses. In addition, 75 percent of the Highways Agency's 82 percent may be paid to the utility company in advance by lump sum or installment. State DOTs should consider such a cost-sharing arrangement for utilities located on public and private rights-of-way. Theoretically, this would be an incentive for utility companies to relocate facilities in a timely manner. It also would eliminate costly, time-consuming arguments over who has prior rights. In addition, State DOTs should consider paying preliminary engineering costs up front in exchange for an agreement from utilities to complete work in a timely manner. They also should consider paying construction costs up front under a pre-financing agreement in cases where disagreements over project costs threaten delays.

Acquisition of Right-of-Way for Utilities—Several European countries visited acquire right-of-way for utility purposes. Some State DOTs in the United States do this also. When State DOTs are in the process of acquiring right-of-way, they could consider acquiring, in consultation with affected utilities, sufficient right-of-way to accommodate utility needs. This would minimize inconvenience to property owners created when both DOT and utility representatives approach them to acquire property rights.

Damage Prevention—Damage to underground utilities by excavation activities is a problem in Europe. Utility companies in Germany are responsible for identifying their underground facilities and providing this information to highway contractors before excavation. Highway contractors in the Netherlands are required to call a national information center to obtain information about underground utilities before they begin excavation activities. Highway contractors in England must notify all affected utilities before they begin to dig. Despite these activities, damage to underground utilities continues to occur. The same is true in the United States, where extensive one-call notification programs have been developed to reduce damage to underground utilities caused by excavation activities. To protect underground utilities from unnecessary damage, State DOTs should use one-call notification centers at an appropriate level of participation and provide sufficient oversight to assure that highway contractors fully participate in one-call notification programs.

Protected Highway Designation—In England, the Highways Agency has the power to designate a roadway as protected, precluding new utility installations. Existing utilities on newly designated roads may repair facilities and make service connections, but may not expand or replace facilities on the right-of-way. This designation applies to motorways and other major highways. As DOTs in the United States search for ways to increase highway capacity and facilitate traffic flow on the National Highway System, consideration should be given to ways in which this idea may apply.

Minimizing Pavement Cuts—Pavement cuts are a significant problem in Europe. In Germany, underground utility crossings on major roads are made by boring, jacking, or directional drilling, but pavements frequently are cut for utility crossings on lesser roads and for fiber optics installations on streets. Pavement cuts are a problem in the United States also. Pavements on lesser rural highways are cut routinely for utility crossings, while in cities pavements frequently are cut to access utilities located longitudinally beneath streets. Fiber optics installations are becoming particularly troublesome as streets are torn up for installations and then poorly repaired. Efforts need to be made in the United States to use non-destructive technologies for highway and street crossings and to better control the frequency of pavement cuts to access or install utilities under streets.

Geographic Information Systems—GIS is used in Europe for mapping right-of-way properties. In Norway and England, software programs have been developed and GIS is used extensively. Similar activities are under way in the United States, including efforts to use GIS for utilities, but more research on the feasibility of using GIS to map utilities should be initiated.

Accommodation of Fiber Optics and Wireless Telecommunications—Fiber optics and wireless telecommunications are being installed in Europe. A fiber optics installation on highway right-of-way is being considered in Norway, although there are no plans to receive cash or services in exchange for allowing this installation. In Germany, cables used by mobile phone operators have been installed in tunnels for a fee, and wireless telecommunications towers have been installed in privately owned service areas along motorways. Procedures are being developed in the Netherlands to accommodate fiber optics and wireless telecommunications towers. Compensation has not been seriously considered, but provisions will be made just in case. Consideration is being given in England to accommodating fiber optics and wireless telecommunications towers in exchange for cash or services.

Many States in the United States have entered into resource-sharing arrangements with both fiber optics and wireless telecommunications companies. States have received cash and services in exchange for use of highway right-of-way. Intelligent Transportation Systems have been enhanced greatly by these installations. State DOTs should continue pursuing resource-sharing arrangements, but care should be taken to assure that the safety, operations, and integrity of highways are not compromised.

PROJECT DEVELOPMENT

Right-of-Way and Utilities Functions in Design-Build Process

The team supports FHWA and AASHTO efforts to examine the feasibility of incorporating right-of-way functions, as well as utilities, into the design-build process. The team encourages State right-of-way and utilities personnel to study the benefits of design-build contracting, including shortening the project development process by eliminating many procedural procurement processes.

Corridor Preservation

The team recommends establishment of a work group through FHWA to reevaluate methods for corridor preservation using the "1990 Report of the AASHTO Task Force on

Corridor Preservation" as a starting point. The team suggests creation of one or more pilot projects to test corridor preservation and land consolidation techniques. The team recommends evaluating adverse effects of not doing early right-of-way acquisition, including increased costs because of value appreciation and adverse impacts on property owners forced to hold property they cannot sell.

This should be a joint effort with AASHTO subcommittees responsible for statewide transportation planning, land use and environment, and right-of-way. As background, England's Blight Acquisition Program offers relief to property owners adversely affected by a pending project. This program provides for early acquisition of such properties.

Rights of Entry and Early Acquisition Methods

The team recommends that FHWA and States evaluate methods used in various countries for rights of entry and early acquisition to facilitate early entry onto property for project construction. They should consider expanding these methods by using risk management concepts, while ensuring that property owner rights are protected.

Information Clearinghouse on Right-of-Way and Utilities Databases

The team encourages the AASHTO Right-of-Way and Utilities Subcommittee to establish an information clearinghouse on right-of-way and utilities databases, including GIS, for project development, tracking, and management.

Chapter Eight
IMPLEMENTATION ACTIVITIES

 RIGHT-OF-WAY PROGRAM

After the March 2000 scanning study, the rest of the year was devoted to:

- Formulating an FHWA policy on land consolidation, which was issued in December 2000.

- Conceptualizing possible experimental right-of-way projects to test and evaluate specific European right-of-way practices.

- Preparing an application for scanning implementation funding.

This chapter describes implementation activities under way based on the Right-of-Way and Utilities scanning team's recommendations.

Land Consolidation

Land consolidation is used extensively in Norway and Germany. In both countries, the process can be judicially enforced if it is found to be in the public interest. This concept was considered particularly applicable to the proposed Interstate 69 route in Indiana, Illinois, Tennessee, Arkansas, Mississippi, Michigan, and Texas. This route extends a distance of 1,600 miles, much of which will be built on new alignment. When an indication of interest in land consolidation was solicited from the States along this route, several responded positively.

Staff from the FHWA Office of Real Estate Services (ORES) developed a draft policy, which was circulated for review and comment. They subsequently met and coordinated with the FHWA Office of Human Environment, National Environmental Policy Act Facilitation and Chief Counsel to work out details of the policy and funding and reimbursement procedures. Based on input from the various offices, it was determined that the following guidelines would be applied in developing a land consolidation policy:

- Use of the land consolidation technique will be voluntary and must be legally feasible under State law.

- Funding authority is from the National Environmental Policy Act and it is for implementing regulations to undertake environmental mitigation.

- The environmental assessment phase is the most appropriate time to consider the decision to use the land consolidation technique. This decision will be documented accordingly.

FHWA's policy on land consolidation was issued December 8, 2000. The scanning team recommends that a survey of the use of this policy be undertaken in 2002 and additional marketing activities be considered. In a conference call with the FHWA Mississippi Division office, it was confirmed that Mississippi is actively considering using the land consolidation policy in the I-69 corridor. FHWA has extended the policy for use in all States as appropriate.

CHAPTER 8

Right-of-Way Experimental Projects

The four State right-of-way managers who participated in the scanning study were interested in European right-of-way practices aimed at streamlining or expediting the right-of-way process. Using many of these practices in the United States would require changing Federal laws or regulations. To facilitate a test and evaluate formats for such initiatives, ORES staff drafted conceptual proposals for consideration. Funding for administrative costs related to implementation initiatives was made available in late 2000.

Following the scan, several reports were prepared:

- "Summary of International Scanning Program for Right-of-Way and Utilities" was prepared on the final day of the scanning study in London, England.

- Action proposals were prepared and distributed at the State Right-of-Way Director's meeting in Savannah, Georgia, in May 2000.

- An agenda was developed for the first Scanning Implementation Task Group (SITG) meeting on January 8, 2001.

The reports and agenda describe implementation initiatives. In most cases, ORES prepared a concept statement and shared it with SITG members. All of the initiatives were discussed at the first SITG meeting with the understanding that the scan proposals would be formulated and then presented, discussed, and considered by FHWA at the second SITG meeting on March 22, 2001. The agenda for that meeting outlined the discussion and consideration of implementation on the following experimental projects.

Appraisal Review Modification—Conceptual proposals were received from Florida, Michigan, Wisconsin, and Washington. These experimental projects to evaluate eliminating appraisal review requirements under certain circumstances have been approved by FHWA and are all under way.

Appraisal Waivers exceeding $10,000—Experiments in Florida and North Carolina were approved at $20,000. A proposal from Michigan is pending FHWA approval.

Incentive payments—Florida and Virginia are implementing FHWA-approved projects. A report on the benefits and cost savings of the Virginia project is included in this report in Appendix D. A previous project in Michigan had excellent results.

Same Person Appraising and Negotiating On Properties Exceeding $10,000—California has an experiment under way extending the conflict of interest limit to $25,000. Each California DOT district is authorized to adopt this procedure on one project.

Appraisal/Replacement Housing Payment Calculation—Colorado and Arizona may consider this one-step process for experimentation, although FHWA has received no State proposals. FHWA may need to do a broader solicitation for a State experiment to be undertaken in this area.

CHAPTER 8

Right-of-Way Exchanges, Meetings, and Presentations

The Right-of-Way and Utilities Scanning Study has generated numerous follow-up activities, consultations, and meeting presentations by representatives of the European host delegations.

- A representative of England's Highways Agency made a presentation at the FHWA/AASHTO Right-of-Way and Utilities National Conference in 2000.
- Representatives from the Netherlands made presentations at the 2000 and 2001 FHWA/AASHTO Right-of-Way conferences.
- A representative of England's Valuation Agency Office was consulted for advice on how the office handles relocation reestablishment payments in England.
- The representative from England's Valuation Agency Office made a presentation at the IRWA Uniform Act 30th Anniversary Symposium held in November 2001. The presentation covered relocation assistance program benefits and payments in England related to acquisitions under the threat of eminent domain.
- Several scanning team members made presentations to professional organizations and government officials with right-of-way responsibilities.
- As part of a training initiative based on the scan, a symposium was arranged in cooperation with Morgan State University in Baltimore, Maryland, on the development of an undergraduate curriculum on real estate acquisition. Representatives from the University of Oklahoma and Delaware Technical and Community College participated in the symposium.
- The Scan Implementation Task Group met four times in 2001 and participated in the Real Estate Acquisition Curriculum Symposium at Morgan State University in October 2001.

Future Actions

Team members noted several other areas during the scan worthy of consideration for future action in the United States. The FHWA Office of Real Estate Services, with the cooperation of AASHTO, will take the initiative to explore how such techniques may be tested and evaluated. In some cases, such techniques may already be used in part or in total. Areas for exploration include:

- Practices that afford greater property owner input, including property owner involvement before completion of final right-of-way plans and an extensive property owner interview process.
- A practice that allows for a single agent for appraisal and negotiation to limit the number of contacts dealing with property owners.
- Facilitating settlements and payments to property owners by the use of mediation and quick payment processes.
- Greater flexibility to compensate for expenses related to relocation assistance. A national study of business relocation assistance and a pilot project in Rhode Island addressed issues relating to business relocation expense reimbursement. More information has been gathered from representatives of countries visited during the scanning study who have subsequently visited the United States.

This exchange should be continued in the context of efforts to update the Federal Uniform Relocation Act.

- Ways to increase flexibility in the area of early right-of-way acquisition. Activities under way in the United States include:
  - The Center for Transportation and the Environment at North Carolina State University held a videoconference on "Integrating Right-of-Way and Environment for Better Results" on October 16, 2001.
  - An FHWA Planning and Environment Core Business Unit research project entitled "Integration Solutions: Integrating and Streamlining Transportation Development and Decision Making" is being conducted. The project involves the FHWA offices of Planning, Environment, Real Estate, Design, and Construction.

The objective of these efforts is to evaluate existing environmental and right-of-way regulations to determine any changes that should be considered to accommodate early right-of-way acquisitions.

- Alternatives for developing right-of-way training. Several European countries have arrangements with colleges and universities that may be worthwhile to explore in the United States. Initiatives with the following institutions are under way:
  - Morgan State University, Baltimore, Maryland
  - Delaware Technical and Community College, Stanton, Delaware
  - Marylhurst University, Portland, Oregon

Delaware Technical and Community College and Marylhurst University have specific capabilities in the area of distance learning, which may be particularly relevant to the highway right-of-way program.

UTILITIES PROGRAM

After the March 2000 scanning study, the balance of the year was devoted to the following utilities-related areas:

- Conceptualizing possible utilities-related experimental projects to test and evaluate specific innovative European practices.
- Developing possible research and technology transfer projects to obtain and disseminate information about ongoing activities in the United States similar to European utility practices.
- Encouraging State DOTs and utility companies to investigate the feasibility of applying specific innovative European utility practices in the United States.

The following sections detail implementation activities now in progress.

Experimental Utilities Projects

The FHWA Office of Program Administration has initiated the following experimental projects to evaluate innovative European utility practices.

Investigation of the feasibility of paying preliminary engineering costs for all utility relocations—The Virginia DOT began a pilot program in September 2000 in which it agreed to pay 100 percent of all preliminary engineering costs for utility relocations, regardless of who had prior rights. Results to date have been statistically inconclusive, but early returns indicate that benefits outweigh costs. The pilot project is continuing.

Investigation of the feasibility of recognizing pipelines as a mode of transportation— The North Carolina State University Center for Transportation and the Environment initiated a literature search and found that the Texas Transportation Institute has designed a system to use pipelines to carry freight from Dallas, Texas, to Laredo, Mexico. The institute is looking for a funding source to build a prototype.

Utility Research/Technology Transfer Projects

The Office of Program Administration has initiated the following research and technology transfer projects to obtain and disseminate information about ongoing activities in the United States similar to innovative European utility practices.

Investigation of techniques to avoid unnecessary utility relocations—A contract was awarded to Nichols Consulting Engineers, Inc., in May 2001 to investigate and report on innovative techniques for avoiding unnecessary utility relocations. A final report will be distributed to FHWA, State DOTs, and utility organizations.

Investigation of methods to control the frequency of pavement cuts—A contract was awarded to the Transtec Group, Inc., in March 2001. Transtec Group will review and report on workable State and local policies for controlling the frequency of pavement cuts and state-of-the-art trenchless technology methods being used to avoid the need to cut pavements to install or access underground utilities. A final report will be distributed to FHWA, State DOTs, and utility organizations.

Evaluation of feasibility of placing utilities underground to reduce utility pole collisions—The Transportation Research Board (TRB) Utilities Committee (A2A07) convened a task force to look at utility pole safety, investigate countermeasures such as locating utilities underground, and develop a report on state-of-the-art utility safety. The final draft of this report was presented to TRB on January 15, 2002. Copies of the report will be distributed to FHWA, State DOTs, and utility organizations. Follow-up implementation actions are being developed.

Utility Meetings and Presentations

FHWA has encouraged State DOTs and utility companies to investigate the feasibility of applying specific innovative European utility practices in the United States. Efforts have included presentations and workshops at utility-related conferences, such as AASHTO Right-of-Way and Utilities conferences, National Highway/Utility Educational conferences, Mid-Atlantic Right-of-Way and Utilities conferences, and Transportation Research Board conferences. Additional efforts include:

Encourage State DOTs and utility companies to coordinate, cooperate, and communicate—The "AASHTO Guidelines and Best Practices for Utilities" were distributed to FHWA field offices for subsequent distribution to State DOTs on March

27-28, 2001. These guidelines and best practices stress the importance of good coordination, cooperation, and communication, and provide examples of successful State DOT programs. A pilot National Highway Institute highway and utility issues training course was conducted in North Carolina on March 27-28, 2001. Regular presentations began in Illinois on January 22-23, 2002. Numerous other presentations have been scheduled.

The FHWA Office of Asset Management is developing a videotape encouraging design and construction personnel to coordinate, cooperate, and communicate more effectively. This video will be distributed to FHWA, State DOTs, and utility organizations. A memorandum was sent to all FHWA field offices on November 28, 2001, urging them to take the initiative in encouraging State DOTs and utility companies to coordinate, cooperate, and communicate in the development of highway projects.

Encourage use of master utility agreements—The "AASHTO Guidelines and Best Practices for Utilities" distributed to State DOTs encourage the use of master agreements for utility relocations and the inclusion of utilities in design-build contracts.

Future Actions

Several other areas noted during the scan are worthy of consideration. The FHWA Office of Program Administration, with the cooperation of AASHTO, will take the initiative to explore how such techniques may be valuated in the United States. Areas for exploration include:

- Encouraging States to establish utility corridors.
- Determining the feasibility of highway contractors installing utilities.
- Determining the feasibility of mapping utilities using GIS, GPS, and other innovative techniques.

RIGHT-OF-WAY AND UTILITIES GUIDELINES AND BEST PRACTICES REPORT

A draft of AASHTO's "Right-of-Way and Utilities Guidelines and Best Practices" report was issued on January 21, 2000. The Subcommittee on Right-of-Way and Utilities believes it would be beneficial to amend this report to include practices the Right-of-Way and Utilities scanning team observed in Europe that may also be useful in the United States.

Suggested items to add to the report follow.

Project Development

- Use of an interview process with property owners in cases where dialogue and feedback about the project design, schedule, or construction plans may result in an improved situation for all concerned.
- The concept of land consolidation can be used to mitigate impacts caused by property acquisitions that sever large parcels of land, create access problems, and disrupt the economic and social fabric of a community. Land consolidation involves acquiring remainder properties or replacement lands and rearranging land ownerships to operate economically and with reduced social impacts. Using

this technique would be considered in the environmental assessment stage of a project's development. This concept and the authority for land consolidation and its eligibility for Federal funding reimbursement is discussed in a December 8, 2000, FHWA memorandum.

- European project development practices mirror those in use and recommended in this report. These practices ensure there is right-of-way input to project development from all aspects, including planning, environment, design, scheduling, and construction. Our European right-of-way counterparts realize the importance of being involved in this process. They also know that good communication and accountability for all involved in the process are key.

Appraisal and Appraisal Review

As part of scan implementation activities, FHWA, working cooperatively with the States, has approved several experiments in the area of appraisal and appraisal review. These experiments involve increasing limits for appraisal waivers and eliminating appraisal reviews in certain circumstances. The experimental projects are under way and may result in future revisions of FHWA policies or regulations.

Acquisition

As part of scan implementation activities, experiments are under way in several States on using incentive payments and expanding the conflict-of-interest limits that allow the same person appraising property to also negotiate for it. After evaluation, these experimental projects may result in revised FHWA policies and regulations.

Some of the best practices for acquisition contained in the guidelines are being used effectively in Europe. These include:

- Prompt payment and having warrants available at the beginning of negotiations.
- A caseworker or single-agent approach to performing acquisition and relocation activities.
- Use of mediation to resolve right-of-way acquisition disputes.

Training

The report should include a new best practice on cooperating with institutions, such as colleges or universities, to develop both traditional and non-traditional training in the right-of-way and utilities program areas. This is being done successfully in some European countries.

Appendix A
TEAM MEMBERS

Team Members and Affiliations*

Richard Moeller (Co-Chair)
Team Leader, Technical Services Group
Office of Real Estate Services
Federal Highway Administration
1172 Commodore Court, #102
Fort Pierce, Florida 34949
Phone: (561) 429-0064
E-Mail: rmoeller@orcolan.com

Joachim Pestinger (Co-Chair)
Director, Real Estate Services
Washington State Department of
Transportation
PO Box 1227
Oting, Washington 98360
Phone: (360) 893-6617
E-Mail: pestinger@earthlink.net

Myron Frierson
Administrator, Real Estate Division
Michigan Department of Transportation
PO Box 30050
Lansing, Michigan 48909
Phone: (517) 373-2200
E-Mail: friersonm@mdot.state.mi.us

Wayne Kennedy
International Right-of-Way Association
38807 South Starwood Drive
Tucson, Arizona 85739
Phone: (520) 2818-1812
Fax: (520) 818-1813
E-Mail: wayneken1@earthlink.net

John A. Almborg
(Delegation Coordinator)
American Trade Initiatives, Inc.
3 Fairfield Court
Stafford, Virginia 22554-1716
Phone: (540) 228-9700
Fax: (540) 288-9473
E-Mail: j.almborg@gte.net

Catherine Muth
O.R. Colan Associates, Inc.
128 North Street, Suite 201
Bluefield, West Virginia 24701
Phone: (304) 327-6968
E-Mail: muth@orcolan.com

Janet Myers
Director, Right-of-Way Division
Maine Department of Transportation
Center for Transportation and the
Environment
Box 8601
Raleigh, North Carolina 27695-8601
Phone: (919) 515-8041
E-Mail: jlmyers@unity.ncsu.edu

Paul Scott
Highway Engineer
(Utilities Coordination)
Office of Program Administration
Federal Highway Administration
400 Seventh Street SW
Washington, DC 20590
Phone: (202) 366-4104
Fax: (202) 366-3988
E-Mail: paul.scott@fhwa.dot.gov

Stuart Waymack
Director, Right-of-Way and Utilities
Virginia Department of Transportation
1401 East Broad Street
Richmond, Virginia 23219
Phone: (804) 786-2923
Fax: (804) 786-1706
E-Mail: waymack_sa@vdot.state.va.us

Adele McCormick
(Report Facilitator)
Technical Writer
4017 Libby Road NE
Olympia, Washington 98506
Phone: (360) 943-4708
Fax: (360) 943-4708
E-Mail: McAdele@aol.com

---
* Affiliations at time of scanning study. Addresses, phone numbers, and e-mail addresses current at time of publication.

APPENDIX A

TEAM MEMBER BIOGRAPHIES

The following biographies were written before the Right-of-Way and Utilities Scanning Study to provide information about team members to the host delegations.

Richard Moeller is a real estate specialist/manager for FHWA in Washington, D.C. Moeller serves on a leadership team of three senior managers that directs operations of FHWA's Office of Real Estate Services. He is responsible for developing and issuing national policies and procedures related to acquisition of real property for highway right-of-way. Specific responsibilities concern valuation and acquisition of real property through negotiation and condemnation. This process of right-of-way procurement also includes assistance to residential and business occupants displaced by proposed construction activity. Moeller has served FHWA in various capacities in its highway right-of-way program for more than 36 years. He has a bachelor's degree in business administration from the University of Iowa. He is secretary of the AASHTO Subcommittee on Right-of-Way and Utilities.

Joachim Pestinger is director of real estate services for the Washington State DOT in Olympia, Washington, where he administers acquisition, appraisal, relocation assistance, title clearance, and property management activities. He approves up to 900 property purchases, mediated settlements, or stipulated judgments a year. He establishes policies and procedures for the department and other State and local public agencies. After graduating from Brigham Young University, he became an appraiser and manager for Clark County in Washington and then property supervisor for Seattle, Washington. He holds the senior member designation of the International Right-of-Way Association, and has taught courses in appraisal, negotiation, engineering, land titles, court testimony, real estate law, and property management throughout the United States and Canada. He was lead author of Course #700, Introduction to Property Management, and Course #801, Land Titles. He serves on the eminent domain faculty of Law Seminars International, Inc., and has been a speaker at appraisal conventions and education conferences for the Washington State attorney general's office. As an AASHTO member, he represents 17 western States on the executive board of the Subcommittee on Right-of-Way and Utilities. He also is on AASHTO's Special Committee on International Activity Coordination.

Adele McCormick is a technical writer for the Washington State DOT in Olympia, Washington. She is responsible for writing and editing reports, manuals, and studies developed by the Olympia Service Center Design Office and other Washington State DOT offices and teams. She is web master for several Washington State DOT web pages covering roadside and site development, design-build, real estate acquisition, and real estate asset management. McCormick has served as technical writer for numerous Washington State DOT value engineering studies and process improvement teams. She recently completed two comprehensive process improvement efforts covering real estate acquisition and real estate asset management. She has a bachelor's degree in speech from Washington State University.

Myron Frierson is division administrator for the Michigan DOT in Lansing, Michigan. Frierson directs the department's Real Estate Division, which provides direction and develops right-of-way policies and procedures for statewide right-of-way operations. His duties include providing administrative direction to all phases of right-of-way activities, such as appraisal, acquisition, relocation, and property management. He administers the Michigan DOT's oversize and overweight vehicle permitting, billboard permitting, and utility coordination programs. In this capacity, he has emphasized cross training of staff and the generalist property analyst concpet. He has implemented several process improvements that have resulted in processes that are more responsive to customer needs. In his 18 years with Michigan DOT, he has held administrative positions in highway district administration and financial management. Frierson has a bachelor's degree in accounting from Michigan State University and is a certified public accountant. He is a member of the AASHTO Subcommittee on Right-of-Way and Utilities.

Wayne Kennedy is international president-elect of the International Right-of-Way Association (IRWA), with headquarters in Gardena, California. He is responsible for the committees on Ethics, Asset Management, Local Public Agency, Liaison, Pipeline, Relocation Assistance, Environment, Surveying, Transportation, Utilities, Valuation, and Professional Development. He has served as director of FHWA's Office of Right-of-Way, director of right-of-way for New Mexico, manager of appraisal and appraisal review in Florida, appraiser for the U.S. Army Corps of Engineers, and has spent 34 years at FHWA. He has bachelor's and master's degrees in business administration with a real estate major from San Jose State University in California. He also holds certificates in real estate and public administration from the University of California. He holds IRWA's senior designation and the American Society of Appraisers' senior and master governmental appraiser designations.

Catherine Colan Muth is president, chief executive officer and owner of O.R. Colan Associates, Inc., a Bluefield, West Virginia, consulting firm specializing in providing all phases of right-of-way acquisition between design and construction of public works projects. Her company's clients include departments of transportation, airport authorities, and local public agencies throughout the United States. She is lead author for a Transportation Research Board study on "Innovative Practices to Reduce Delivery Time for Right-of-Way in Project Development." After graduating from West Virginia University with a bachelor's degree in political science and English, she completed two years at Bluefield State College in West Virginia studying computer science and accounting. She is a member of IRWA and the Environmental Assessment Association. She was nominated as West Virginia's entrepreneur of the year in 1996.

Janet Myers is director of the Right-of-Way Division of the Bureau of Project Development for the Maine DOT in Augusta, Maine. She is responsible for statewide policies, procedures, and operations for property valuation, acquisition, management, and relocation activities, as well as utilities, accommodation, and relocation functions. Maine DOT is pursuing significant program expansion, which calls for evaluating and streamlining many of the standards and processes relating to right-of-way and utilities activities. Myers is an attorney with a background in real estate and environmental law. Before joining the Right-of-Way Division, she was a trial attorney and major projects manager for Maine DOT. Myers holds a bachelor's degree in history from Stanford University and a law degree from Boston University School of Law. She is a member of AASHTO's Subcommittee on Right-of-Way and Utilities and she serves as a research panel member on NCHRP 25-23, "Environmental Information Management and Decision Support System for Transportation."

APPENDIX A

Paul Scott is a highway engineer for FHWA in Washington, D.C. Scott coordinates the relocation and accommodation of utilities on Federal-aid highway projects. His work includes wireless telecommunication towers, subsurface utility engineering, underground damage prevention programs, and utility pole crash reduction programs. Before taking on his utilities responsibilities in 1989, he served the FHWA for 20 years in a number of highway engineering capacities. Scott has a bachelor's degree in civil engineering from the University of Tennessee. He is a licensed professional engineer and serves on technical committees of the American Society of Civil Engineers, IRWA and TRB.

Stuart Waymack is state director of the Right-of-Way and Utilities Division of the Virginia DOT in Richmond, Virginia. He is responsible for statewide acquisition of all real property and relocation of families, businesses, and utilities in the path of or affected by transportation improvement projects in Virginia. He is also responsible for developing strategies for implementing new Federal and State laws that affect the transfer of private lands to the Commonwealth of Virginia for transportation projects. Waymack has more than 40 years of service with the Virginia DOT and has served as an appraiser, negotiator, district right-of-way manager, and assistant state right-of-way engineer. He attended the University of Richmond in Virginia and has been active in several local and national right-of-way and appraisal organizations. Most recently, he has been a member of an FHWA task force on the installation of fiber optics on the U.S. interstate system under the shared-resource concept.

Appendix B
AMPLIFYING QUESTIONS

Several months before the Right-of-Way and Utilities Scanning Study, team members developed the following list of topics and questions. These amplifying questions were submitted to the host countries to give them an indication of the basis and scope of information the team desired. The host countries used these questions as a guide for developing their presentations.

I. Right-of-Way and Utility Laws, Regulations, Policies, and Programs

   A. Please provide an overview of laws on private property ownership in your country and acquiring private property for highway purposes. Are these laws the same for all levels of government?

   B. What compensation are you required to give a property owner before private property can be taken for public use?

   C. When a property owner will not voluntarily sell a piece of property, what procedures do you follow to acquire it?

   D. What laws govern controlling, managing, and disposing of improvements (primarily buildings) when you acquire a property for highway use?

   E. Please provide an overview of your laws on utilities in highway rights-of-way. Are they different for publicly and privately owned utilities?

   F. Do your laws require compensation for impacts caused by transportation projects, such as noise, business interruption, etc.?

   G. What are the most important right-of-way and utilities challenges facing your agency in the next five years?

II. Right-of-Way and Utility Involvement in Project Development

   A. Please share the practices, policies, and techniques you use to minimize time requirements for delivery of right-of-way for project construction. How do you ensure there is adequate time to perform right-of-way functions?

   B. What role do right-of-way and utilities staff members have during planning, design, and construction of a project?

   C. During project decision-making, how much importance is given to the impact of the project on local neighborhoods and on viability of local businesses?

   D. During project development, when do you begin coordination with utility companies and departments on the use of the highway right-of-way for utilities and the location or relocation of those utilities?

APPENDIX B

    E. What methods do you use to minimize the cost of right-of-way?

    F. What techniques do you use to ensure that your design and right-of-way plans are sufficiently complete to proceed with right-of-way acquisition and utility relocation?

III. Property Appraisal and Appraisal Review

    A. In the United States, an appraisal is used to determine the value of property needed for highway right-of-way. What process do you use to determine how much to pay for properties?

    B. If you use appraisals to determine property values:
        1. What are the requirements for appraisers?
        2. What are the requirements for an acceptable appraisal?
        3. What problems have you encountered with the appraisal process?
        4. What is your process for reviewing appraisals?
        5. Are there appraisal associations and what are their requirements?

    C. When compensating a property owner, what consideration is given for:
        1. Tenant-owned improvements?
        2. Business damages?
        3. Damages to the remainder of a property when only a portion of it is required for highway right-of-way?

IV. Acquisition of Property Rights, Easements, and Permits

    A. What types of property rights do you acquire? Under what circumstances and for what purposes do you acquire them?

    B. Have you developed any effective techniques for resolving disputes related to compensation, such as mediation, arbitration, settlements, etc.?

    C. How do you negotiate to acquire property and what information do you give property owners?

    D. What techniques have you found to expedite acquisition of property needed for right-of-way?

V. Relocation Assistance to Owners, Tenants, Businesses, and Farm Operations

    A. What relocation assistance do you provide to owners, tenants, businesses, nonprofit organizations, and farm operations when their property is acquired and they must move to a new location?

    B. What problems do you encounter in your relocation process?

VI. Utility Coordination, Adjustments, and Relocation for Highway Projects

A. Are utility companies privately or publicly owned? Do you have different standards for each?

B. Please give an overview of the requirements for utility companies when placing and relocating utilities in the highway right-of-way.

C. What responsibilities does the highway agency have for coordination and compensation when utilities are placed or relocated in the highway right-of-way?

D. What records does your agency maintain on utility installations in the highway right-of-way? Is subsurface utility engineering used to show utility locations?

E. Do you have written agreements between your highway agency and utility companies to describe their respective responsibilities for financing and accomplishing relocation and adjustment work?

F. How do you resolve conflicts over utility work and relocation, including scheduling, cost, location, and right-of-way acquisition?

G. What permits are required for location of utilities in the right-of-way and how are they enforced?

H. What are your standards for aboveground and underground utility installations? What considerations are used in establishing standards?

VII. Public Involvement

A. Do you have legal requirements for advising the public of proposed highway projects?

B. Please give an overview of your process for advising the public and receiving public input about proposed highway projects.

VIII. Property Management of Real Estate Acquired for Highway Right-of-Way

A. Please describe your process for disposing of property you own that is no longer needed (excess property).

B. What records are kept on excess property? Are they kept electronically? What software is used?

C. Do you lease excess property? If so, under what conditions and to whom do you lease it, and how is it valued?

D. Do you lease airspace over or along the highway right-of-way? What laws and regulations do you have on airspace use?

E. Are fiber optics and wireless telecommunication towers accommodated on the highway right-of-way?

   F. Does your country protect roadway capacity and safety by controlling access to the roadway system? How and where do you control it?

IX. Training Programs and Mentoring Procedures for Right-of-Way Staff

   A. Please describe the education and training required for right-of-way employees.

   B. What training and/or mentoring programs do you provide for your employees? Does your agency pay for required training?

   C. Do you require any proficiency certifications, examinations, licenses, or other standardized measures of competency?

   D. Do you use any training methods you think are especially effective?

   E. How do you measure the effectiveness of training?

   F. What role does technology play in:
      1. Training?
      2. Project management?
      3. Property management?

APPENDIX C
CONTACTS IN HOST COUNTRIES

The following list includes people who served as points of contact for the Right-of-Way and Utilities Scanning Study. The team would like to express its gratitude to each of them for their contributions to the success of the trip.

NORWAY

Roar Midtbo Jensen
Norwegian Public Roads Administration (NPRA)
Grensevegen 92
Helsfyr, Oslo
NORWAY

GERMANY

Hans Mundry
Federal Ministry of Transport, Building and Housing
Robert-Schuman-Platz 1
D-53175 Bonn-Bad Godesberg
GERMANY

THE NETHERLANDS

Paul van der Kroon
Ministry of Transport, Public Works and Water Management
Johan de Wittlaan 3
P.O. Box 20906
2500 EX The Hague
THE NETHERLANDS

Hans J.P. van Douwe
Ministry of Transport, Public Works and Water Management
Johan de Wittlaan 3
P.O. Box 20906
2500 EX The Hague
THE NETHERLANDS

UNITED KINGDOM

James Bradley
The Highways Agency
St. Christopher House
Southwark Street
London SE1 OTE
UNITED KINGDOM

Malcolm Macleod
The Highways Agency
St. Christopher House
Southwark Street
London SE1 OTE
UNITED KINGDOM

John Powell
The Highways Agency
St. Christopher House
Southwark Street
London SE1 OTE
UNITED KINGDOM

APPENDIX C

Listed below are the individuals the team met with during the scanning study. The team members wish to express their sincere gratitude to these individuals for their time and hospitality and the valuable information they provided.

NORWAY

Norwegian Public Roads Administration
Per Arne Andresen
Olaf Ballangrud
Anders Hagerup
Tor Hoie
Roar Midtbo Jensen
Dagfinn Loyland
Tormod Olsen
Ola Omenas
Stein Rinholm
Eric Westerlund

The Agricultural University of Norway
Professor Hans Sevatdal

GERMANY

Bundesministerium fur Verkehr (Federal Ministry of Transport)
Vera Gintzel
Karl-Heinz Johnen
Hans Mundry
Karl F. Ribbeck
Hans Stumpel
Peter J. Weitershagen
Edgar Wittmann

THE NETHERLANDS

Ministerie van Verkeer en Waterstaat (Ministry of Transport, Public Works, and Water Management)
Martien Beemsterboer
E.J. M. Coenen
J.J. Dressing
Henk Gieerveld
A.J J. Hekker
Pieter Jansen
Peter Kieft
E.J.A. van der Boom
Huub van der Kolk
Paul van der Kroon
H.J.P. van Doewe
Gerit van Kekem
Robert J.J. van Winden

ENGLAND

Highways Agency
Mike Ainsworth
Terry Brackenbury
James Bradley
Martin Hobbs
Malcolm Macleod
Mary Moore
John Robinson
John Sherwood

Valuation Office
David Russell-Smith

Appendix D
Virginia Department of Transportation
Woodrow Wilson Bridge Project Report

APPENDIX D

Virginia Department of Transportation
Woodrow Wilson Bridge Project

COST & SCHEDULE SAVINGS FROM THE EARLY MOVE INCENTIVE PROGRAM FOR THE HUNTING TOWER AND TERRACE BUILDINGS

VDOT Right-of-Way Group

December 6, 2001

APPENDIX D

INTRODUCTION

When the design for the new Woodrow Wilson Bridge was approved, one of the biggest challenges to the Virginia Department of Transportation (VDOT) was the relocation of tenants from one Hunting Tower and three Hunting Terrace apartment buildings that are in the path of the new roadway alignment. To maintain the project schedule, the VDOT right-of-way team used an experimental tenant relocation incentive program in cooperation with the Federal Highway Administration (FHWA) to vacate the properties in time for subsequent construction activities to continue as planned. With the recent successful completion of the relocation effort, this paper summarizes the schedule and cost savings achieved through the relocation incentive program. This incentive is a result of FHWA's right-of-way scanning study in Europe in 2000.

BACKGROUND

During the design phase of the Woodrow Wilson Bridge Project, VDOT determined that one tower apartment and three garden apartment buildings had to be vacated to make room for the expanded Interstate 495/95 Capital Beltway immediately adjacent to the new river crossing. The acquisition and demolition of these properties required the relocation of approximately 333 residential units. A right-of-way production team was hired to assist VDOT in the relocation process and given an aggressive schedule for relocating displaced tenants.

Under the original project schedule, goodwill contacts were to begin on June 1, 2000, and all tenants were to have offers for replacement housing payments (RHPs) presented by September 1, 2000. Between 25 and 30 RHP offers needed to be made each month to maintain the project schedule.

Because of various political and environmental factors, VDOT was unable to maintain the original schedule for acquiring the properties and relocating the residents. As a result, offers were presented to the property owners on October 27, 2000, and a public tenant meeting was held on November 6, 2000. The development and approval of a property management plan was required, which resulted in additional delays from the original schedule. On April 2, 2001, VDOT acquired the property and the production company began to make RHP offers to the tenants. As a result, the original 15 months was compressed to eight months, maintaining the overall project schedule. The new schedule required that approximately 42 RHP offers be made each month.

To aid in meeting this compressed schedule, VDOT introduced the Early Move Incentive Program for the affected tenants. The program stated that any tenant in residence on April 2, 2001, and moved within 30 days of receiving an RHP offer would be entitled to a $4,000 incentive. If the resident chose to move between 31 and 60 days of receiving an RHP offer, he would be entitled to a $2,000 incentive. VDOT stated that the program was voluntary and was not subject to negotiation. The incentive program was in addition to relocation assistance benefits due to those displaced. To ensure that the information was transmitted to all residents, VDOT delivered announcements individually to all occupants in the affected properties. Out of the 333 residential units affected, 298 were eligible for the incentive.

APPENDIX D

RESPONSE TO INCENTIVE

The response to the Early Move Incentive Program was outstanding, and the production company has moved all of the displaced residents in eight months. A total of 262 units (88 percent of eligible tenants) have claimed the maximum allowable amount of $4,000. Another 15 units (5 percent of eligible tenants) claimed $2,000. VDOT is working with the remaining 21 residents to process their incentive payments.

LESSON LEARNED

One lesson learned from the program is that if you give people a good reason (i.e., incentive payment), you can motivate them to get things done in your time frame. Thus, many of those displaced elected to move to be assured that they could claim the incentive. Others elected to move at the end of their lease term, and some moved before receiving an RHP offer. The schedule for making RHP computations and offers had to be altered to accommodate individuals who wanted to move immediately. Personnel resources had to be reallocated to accomplish this, but the right-of-way team met the compressed schedule in clearing the right-of-way in an extremely short time period.

INCENTIVE PROGRAM COSTS AND RESULTANT SAVINGS

In total, VDOT expects to pay $1,162,000 to individuals displaced from the Hunting Tower and Hunting Terrace apartments. This includes $1,078,000 paid to date and an additional $84,000 for the remaining claims to be processed.

While this incentive program may at first seem expensive for VDOT, savings in terms of overall project cost and schedule are profound. The incentive program saved VDOT a little more than $6 million from the effects of a seven-month delay. If the relocation effort had taken the full 15 months, VDOT would have had to delay demolition of the properties and the start of construction for the U.S. Route 1 tie-in, advance structures, and main contracts.

Taking into account the added costs related to the incentive program, VDOT saved about $4.8 million by using the Early Move Incentive Program. In addition to these savings, VDOT reduced overhead costs required to manage the condemned properties for an additional seven months.

CONCLUSION

The VDOT Early Move Incentive Program successfully relocated residents from 333 apartment units in seven months—less time than originally scheduled. Even though the program cost the project about $1.2 million, VDOT was able to derive the following direct savings from the accelerated relocations:

- Construction schedule-related savings of about $6 million.
- Reduced property management overhead costs for the condemned properties during the relocation period.
- Met the overall project schedule such that the relocation effort will not have an impact on the opening of the new Woodrow Wilson Bridge. If the relocations were delayed and as a result the opening of the new bridge was delayed, VDOT could

have been assessed a $50,000-a-day penalty by the State of Maryland until the new span was accessible.

In addition to the direct savings VDOT achieved, following are indirect benefits to citizens as a result of the incentive program:

- The program developed goodwill between the tenants and VDOT, which serves as a model for future resident relocation programs.
- Virginians and all users of the Capital Beltway will enjoy reduced commuting expenses and travel times through earlier project completion.

www.ingramcontent.com/pod-product-compliance
Lightning Source LLC
Chambersburg PA
CBHW081843170526
45167CB00007B/2889